MW00514528

Food Security, Gender and Resilience

Through the integration of gender analysis into resilience thinking, this book shares field-based research insights from a collaborative, integrated project aimed at improving food security in subsistence and smallholder agricultural systems.

The scope of the book is both local and multi-scalar. The gendered resilience framework, illustrated here with detailed case studies from semi-arid Kenya, is shown to be suitable for use in analysis in other geographic regions and across disciplines. The book examines the importance of gender equity to the strengthening of socio-ecological resilience. Case studies reflect multidisciplinary perspectives and focus on a range of issues, from microfinance to informal seed systems.

The book's gender perspective also incorporates consideration of age or generational relations and cultural dimensions in order to embrace the complexity of existing socio-economic realities in rural farming communities. The issue of succession of farmland has become a general concern, both to farmers and to researchers focused on building resilient farming systems. Building resilience here is shown to involve strengthening households' and communities' overall livelihood capabilities in the face of ongoing climate change, global market volatility, and political instability.

Leigh Brownhill teaches at Athabasca University, Canada, and is an independent scholar focused on gender, agriculture, and environment.

Esther M. Njuguna is a Scientist focused on gender research for the CGIAR Research Program (CRP) on Grain Legumes at the International Crops Research Institute for Semi-Arid Tropics (ICRISAT), Kenya.

Kimberly L. Bothi is the Associate Director for science and engineering at the Institute for Global Studies/College of Engineering of the University of Delaware, USA.

Bernard Pelletier is a Research Associate at McGill University, Canada.

Lutta W. Muhammad is a Senior Researcher at the Kenya Agricultural and Livestock Research Organisation (KALRO), Kenya.

Gordon M. Hickey is an Associate Professor and William Dawson Scholar in the Department of Natural Resource Sciences, McGill University, Canada.

Other books in the Earthscan Food and Agriculture Series

Competition and Efficiency in International Food Supply Chains
Improving food security
John Williams

Organic Agriculture for Sustainable Livelihoods
Edited by Niels Halberg and Adrian Muller

The Politics of Land and Food Scarcity
*Paolo De Castro, Felice Adinolfi, Fabian Capitanio, Salvatore Di Falco
and Angelo Di Mambro*

Principles of Sustainable Aquaculture
Promoting social, economic and environmental resilience
Stuart Bunting

Reclaiming Food Security
Michael S. Carolan

Food Policy in the United States
An introduction
Parke Wilde

Precision Agriculture for Sustainability and Environmental Protection
Edited by Margaret A. Oliver, Thomas F.A. Bishop and Ben P. Marchant

Agricultural Supply Chains and the Management of Price Risk
John Williams

The Neoliberal Regime in the Agri-Food Sector
Crisis, resilience and restructuring
Edited by Steven Wolf and Alessandro Bonanno

For further details please visit the series page on the Routledge website:
http://www.routledge.com/books/series/ECEFA/

Food Security, Gender and Resilience

Improving smallholder and
subsistence farming

Edited by
Leigh Brownhill, Esther M. Njuguna,
Kimberly L. Bothi, Bernard Pelletier,
Lutta W. Muhammad and Gordon M.
Hickey

Routledge
Taylor & Francis Group
LONDON AND NEW YORK

earthscan
from Routledge

First published 2016
by Routledge
2 Park Square, Milton Park, Abingdon, Oxon OX14 4RN

and by Routledge
711 Third Avenue, New York, NY 10017

Routledge is an imprint of the Taylor & Francis Group, an Informa business

© 2016 Leigh Brownhill, Esther M. Njuguna, Kimberly L. Bothi, Bernard Pelletier,
Lutta W. Muhammad and Gordon M. Hickey, selection and editorial material; individual chapters, the
contributors

British Library Cataloguing-in-Publication Data
A catalogue record for this book is available from the British Library

Library of Congress Cataloging in Publication Data

A catalog record for this title has been applied for

ISBN: 978-1-138-81694-7 (hbk)
ISBN: 978-1-315-74585-5 (ebk)

Typeset in Bembo
by diacriTech, Chennai

Printed and bound by CPI Group (UK) Ltd, Croydon, CR0 4YY

Contents

Foreword

'Resilience' is now firmly in the mainstream of research and practice of international development. The journey of this idea from a fringe concept that was used to describe approaches to managing responses to specific shocks to one that now refers to an alternative paradigm of sustainable development itself, is complete. This evolution has been spurred by a massive increase in the interest of the international donor community in funding initiatives to operationalise the concept for improving the lives and livelihoods of vulnerable communities across the world. Unfortunately, the speed with which resilience has been taken up into practice has also meant that a number of key foundational issues remain unresolved. The manner in which the concept, with its origins in the natural sciences and the study of ecosystems, has been applied to systems with a strong social dimension has meant that a negotiation of issues vital to development progress such as politics, power and culture do not yet have the centrality they deserve within this paradigm.

This is precisely why I was thrilled when the editors of this volume approached me two years ago with the concept of this book. In my opinion, the chapters that follow contribute substantially to our understanding of the manner in which resilience can be deployed to solve some of the most intractable problems of our time. There are at least three clear reasons for this assertion.

First, an understanding of the manner in which resilience can inform enterprises of human development has veered towards the conceptual rather than the practical. Treatise focussed on exploring how the tenets of resilience thinking can apply to food security, climate change adaptation, social protection and disaster risk management borrows heavily from socio-ecological systems (SES) theory. Resilience in the SES is understood through complex ideas such as Panarchy and the Adaptive Cycle and tenets such as functional diversity, iterative learning, and nonequilibrium dynamics. These make for elegant heuristics but do not easily provide practical insights into 'doing development differently.' The chapters in this volume bridge this vital gap by drawing on the complex principles of resilience theory and demonstrating their operational potential through the provision of empirical examples. For example, Chapter 9, "Accountability and citizen participation in devolved agricultural policy-making: Insights from Makueni County, Kenya" considers changes in food security-related policy-making in Kenya after the national adoption of a devolved governance framework in 2010. The authors of the chapter distil their pithy analytical insights into a synthesis of the manner in which public participation in devolved decision-making tangibly supports the operationalization of the tenets of resilience thinking. For instance, they argue that the principle of 'diversity' (central to C. S. "Buzz" Holling's original treatise on resilience) can be brought to life through devolved/participatory decision-making. They argue that including a multiplicity of voices in decision-making processes on food security ensures that the priorities of the marginalised are considered and that possible trade-offs in the resilience for different groups are successfully negotiated. Similarly, Chapter 5 on gendered food- and seed-producing traditions argues that supporting the traditional matrilineal transfer of knowledge about field management, seed selection, grain storage, and food-processing is vital for sustained 'learning' (another key tenet of resilience thinking) as these practices stem from a rich and contextually relevant base of experience on ensuring food security in the face of acute stress.

Second, apart from breathing life into the lofty tenets of resilience thinking, this volume also represents a unique and analytically exciting attempt at bridging three fields of research and practice which include agricultural development, gender relations, and socio-ecological resilience. If the concept of resilience is to tangibly improve the lives and livelihoods of vulnerable communities across the world, it must be effectively coupled with other fields of praxis and thinking to ensure that existing enterprises of development deliver resilient outcomes. The introduction to this volume accurately points out that "given the importance of women in agriculture and development, surprisingly little attention has been given, in East Africa and elsewhere, to the question of the connection between gender equity and socio-ecological resilience within farming systems," and the analysis presented across the chapters that follow successfully brings these previously isolated domains together. For example, Chapter 4, "Land to feed my grandchildren: Grandmothers' challenge in accessing land resources in semi-arid Kenya," makes a strong case for diversifying and strengthening women's access to land. The authors argue that this strategy not only serves to further gender empowerment, it is also a proven means of stemming hunger and malnutrition in the face of climate change. Similarly, Chapter 6, "Banking on change: An ethnographic exploration into rural finance as a gendered resilience practice among smallholders" reviews rural sources of credit to smallholder farmers to examine the gendered aspects of borrowing and lending. The chapter effectively demonstrates that the provision of credit to women strengthens the resilience of households through enhanced food security. As such, the chapter argues for accommodating a sharp understanding of gender dynamics in finance to smallholders because this has the ability to empower women, strengthen agricultural practice, and enhance the resilience of vulnerable households.

Third, apart from its practical explorations of dense resilience concepts and the fact that it bridges previously isolated domains of research and practice, this volume is also important because of its sharp focus on the relationship between socio-ecological resilience and gender. An intuitive and normative understanding of the need to empower women through enterprises of international development is now widely prevalent. Yet robust evidence on the manner in which gender empowerment can support resilience to shocks and stresses, as well the ways in which building resilience can be an effective route to gender empowerment, is still missing. This volume makes considerable progress in filling this gap. The book demonstrates how empowering women through access to land and credit can effectively support the enhanced resilience of households to food shocks across Kenya (as seen in the analysis presented in Chapters 3 and 6). At the same time, the book argues that enterprises of resilience that overlook critically important gender dynamics can be self-defeating and can potentially fail in their mission to ensure that communities thrive despite shocks and stresses. This, for instance, is the implicit logic of Chapter 5, which makes a strong case for acknowledging and incorporating traditional matrilineal food- and seed-producing traditions found in Tharaka, Kenya, in any approach to enhancing the food security in the region. Additionally, the concluding chapter of the volume ties the empirical analysis of the preceding chapters together through the postulation of a novel theoretical construct of the 'resilience umbrella.' This permits researchers and policy makers to analyse gendered dynamics across a complex farming system. This in turn will support the operationalisation of a vision of resilience where gender empowerment is front and centre. In this way, this volume highlights the vital importance of locating empowerment at the centre of resilience, thereby successfully breaking out of the historical tendency of researchers to provide technocratic and scientistic interpretations of the concept.

These are the main reasons for my confidence in recommending this volume to anyone interested in exploring the pathways of food security, sustainable livelihoods, and achieving gender empowerment in the face of a diverse and increasing range of shocks and stresses. An engaging analytical approach that deploys a sound theoretical lens to test empirical examples and bring a novel vision of 'gendered resilience' to life makes this a truly rewarding read.

Aditya Bahadur, Overseas Development Institute, London

Acknowledgements

The research presented in this book collection was completed as part of a project titled "Enhancing Ecologically Resilient Food Security in the Semi-Arid Midlands of Kenya," led by McGill University and the Kenya Agricultural Research Institute between 2011 and 2014 (principal investigators: Gordon M. Hickey and Lutta W. Muhammad). This work was carried out with the aid of a grant from the International Development Research Centre (IDRC), Ottawa, Canada, and with the financial support of the Government of Canada provided through Foreign Affairs, Trade and Development Canada (DFATD). Additional support for the research of Colleen Eidt (Chapter 3) was made available through an IDRC Doctoral Research Award 2012–2013 and the Margaret A. Gilliam Graduate Fellowship 2013–2014, McGill University and June Po through the Social Sciences and Humanities Research Council (SSHRC) of Canada.

We gratefully acknowledge the field assistants, interpreters, government and private sector colleagues, and especially our farmer participants for contributing to this research throughout the course of the project. The editors would like to sincerely thank all the anonymous reviewers who generously donated their time to reviewing the chapters in this book.

Notes on Authors

Kimberly L. Bothi is the associate director for science and engineering at the Institute for Global Studies/College of Engineering of the University of Delaware, USA. During the *Innovating for Resilient Farming Systems in Semi-Arid Kenya* (INREF) project, she was a postdoctoral fellow with the Department of Natural Resource Sciences, McGill University.

Leigh Brownhill teaches at Athabasca University and is an independent scholar focused on gender, agriculture, and environment. She co-led the Gender Team in the INREF project as a research associate in the Department of Natural Resource Sciences, McGill University.

Zipporah Bukania is a research officer with the Centre for Public Health Research of the Kenya Medical Research Institute. She co-led the Nutrition research stream of the INREF project.

Colleen M. Eidt is a PhD candidate in the Department of Natural Resource Sciences, McGill University.

Gordon M. Hickey is an associate professor and William Dawson Scholar in the Department of Natural Resource Sciences, McGill University. He was one of the two principal investigators of the INREF project.

Carly James completed her MA with the INREF project in the Department of Anthropology, McGill University. Her thesis is titled *Enhancing Livelihood Resilience in Makueni County, Kenya: The Role of Informal Credit in Smallholder Farming*.

Timothy Johns is a professor in the School of Dietetics and Human Nutrition, and in the Department of Plant Science, McGill University. He co-led the nutrition research stream of the INREF project.

Esther Kihoro was a research assistant with the Kenya Agricultural and Livestock Research Organisation (KALRO) during the INREF project, working with the Gender research stream.

Immaculate N. Maina is the acting centre director of KALRO's Food Crops Research Institute in Njoro, Kenya. She led the scaling-up activities of the INREF project.

Patrick Maundu is a research scientist with the National Museums of Kenya. His field of expertise is ethnobotany.

Tony Moturi is a managing partner for Policy Options Kenya, a policy and research consultancy.

Megan Mucioki completed her PhD studies in 2014 with the INREF project in the Department of Plant Science at McGill University. Her thesis is titled *Building Seed-Sustaining Households: Defining Seed Security through Informal Seed Systems and Intraspecific Diversity on Semi-Arid, Resource Poor Farms in Kenya.*

Samuel Kimathi Mucioki was an INREF field research assistant for KALRO, working in the Tharaka-Nithi County.

Lutta W. Muhammad is a senior researcher at KALRO. He was one of the two principal investigators of the INREF project.

Erick Mungube is the acting centre director at KALRO's Veterinary Science Research Institute at the Muguga North Research Centre. During the INREF project, he was in charge of the Indigenous Chicken stream and oversaw project activities in the Makueni subcounty.

Elizabeth Nambiro is a senior research officer at KALRO's Headquarters, Nairobi. She was involved in the socioeconomic, monitoring, and evaluation team of the INREF project.

Esther M. Njuguna is a scientist of gender research in the CRP on Grain Legumes for the International Crops Research Institute for Semi-Arid Tropics, Nairobi. During the INREF project, she led the Gender research stream, the Participatory Market Systems Development, and project management activities.

Malo Nzioka is a research officer with KALRO. During the INREF project, he was involved with the Indigenous Chicken research stream.

Bernard Pelletier is a research associate at McGill University. He was the INREF project manager for McGill University.

June Y. T. Po is a PhD candidate in the Department of Natural Resources Sciences, McGill University.

Stephanie Shumsky completed her MSc with the INREF project in the Department of Natural Resource Sciences, McGill University. Her thesis is titled *Wild Edible Plants and Their Contribution to Food Security: An Analysis of Household Factors, Access and Policy in the Semi-Arid Midlands of Kenya.*

1 Introduction

Gender, food security and resilience in Kenya

Leigh Brownhill and Esther M. Njuguna

This book presents a multidisciplinary collection of studies undertaken as part of a 3.5-year research-for-development project in Kenya (2011–2014), led by researchers from the Kenya Agricultural and Livestock Research Organization, or KALRO (formerly Kenya Agricultural Research Institute), the Kenya Medical Research Institute (KEMRI), and McGill University. The project, "Enhancing Ecologically Resilient Food Security through Innovative Farming Systems in the Semi-Arid Midlands of Kenya," involved an integrated sociocultural, economic, agronomic, and health-related inquiry into solutions to the food insecurity and malnutrition crisis in Kenya's eastern counties of Machakos, Makueni, and Tharaka-Nithi. The project deployed an original farmer-led approach to farming system improvement that involved farmer adoption of integrated activities (around the evaluation of varieties of drought-tolerant crops and seed systems, water-harvesting and soil con-servation practices, participatory market systems development, nutrition education, pre- and post-harvest crop handling, and animal health training) in the semi-arid lands directly east of the capital city of Nairobi. Local government officials were involved as key supporters of on-the-ground change by connecting on-farm studies to various networks for wider dissemination of knowledge, and generating work-able recommendations for policy on matters from agronomics and education to health and agricultural services and markets.

Our integrative approach embraced the concept of *gendered resilience*, as it was primarily concerned with understanding and recommending development, policy, and farm-level action to bring about greater gender and generational equity, as an essential part of the building of socio-ecological resilience in semi-arid small-scale and subsistence farming systems.

For the purposes of framing the analysis herein, the term *farming system* incorporates the agricultural and related activities and practices, policies, natural resources, and social relations within a particular agro-ecological zone. Eastern Kenya has had a long and contradictory history of colonial and then global market integration as well as marginalization, at different scales and in various expressions (from terracing campaigns to settlement schemes). Households within these farming systems engage in a range of activities (e.g., mixed livestock and crop farming, local trade, local and migrant waged work) which constitute the practices that shape the overall farming system. In the counties of Machakos, Makueni, and Tharaka-Nithi, the farming system centre consists mainly of small-scale and subsistence farming, aspects of which are described in detail in the chapters that follow.

This book is a collection of articles that represents a cross section of the project's multiple research streams, as it draws together studies in agronomy, natural resource sciences, nutrition, anthropology, sociology, and animal sciences. It is in this way representative of the integrative analysis the project sought to produce, with the examination of diverse research questions through the lens of gendered resilience.[1] The collection concludes with a consideration of the ways in which gendered analyses of food security issues can inform, elaborate, and strengthen conceptual and practical understandings of resilience and inform action at farm, research, and policy levels.

The lines of inquiry pursued by contributors to this collection are situated at the convergence of three fields of scientific thinking and practical action: agricultural development, gender relations, and socio-ecological resilience. As relatively distinctive fields, they have also been historically linked, in particular in Kenya, which has made significant contributions to the field of development, especially development focused on improving the food and nutrition security outcomes of smallholder and subsistence farmers (Bahadur, Ibrahim, & Tanner, 2010; Meinzen-Dick et al., 2011; Quisumbing & McClafferty, 2006; Sanginga, Waters-Bayer, Kaaria, Njuki, & Wettasinha, 2009). The East African region has long provided fertile ground for the emergence and application of gender perspectives in development and research. This history can be traced back to at least the 1950s, when colonial officers' wives formed clubs for African women (Kiswahili: *Maendeleo ya Wanawake*; English: Progress for Women), based on indigenous forms of collective action, and intended to serve as sites for training in literacy, domestic sciences, and agricultural improvement (Wipper, 1975). These clubs were also aimed at drawing women away from the ongoing struggle for independence from Britain, an aim that was largely subverted by those Kenyan women who eventually took over the clubs and formed a nationalist women's organisation after independence in 1963 (Elkins, 2005; Presley, 1992). *Maendeleo ya Wanawake* represented the top-down formation of early women-in-development initiatives that later were adapted worldwide. Its history of struggle between top-down and grassroots perspectives is also reflective of debates in the wider field of women in development, which has progressed through various expressions from women *in* development, to women *and* development, to *gender* and development, to gender equity, empowerment, and *transformation* (Njuki & Miller, 2013; Rathgeber, 1990).

A deep history of African women's autonomous organising—for farming, trade, crafts, and other socio-economic activities—forms the bedrock of success for grassroots women's participatory community development today. The strength of Kenyan women's collective action remains evident in the propensity for women to join and maintain both formal and informal women's groups, savings societies, *merry-go-rounds*, income-generating projects, and wider networks (Brownhill, 2009; Presley, 1992). Poor women, in particular, in both rural and urban areas, are also actively engaged in mixed-gender groups, such as farmer groups and cooperative societies, though they are far from being equally represented in the leadership of many of these groups (e.g., see discussion in Chapters 2 and 3, in this volume).

It is no longer feasible for agricultural extension services to reach individual farmers (Government of Kenya, 2011). Farmer groups and specifically women's groups have been identified by many, with findings reconfirmed in this study, as the unique entry point to rural farming communities for dissemination of agricultural extension messages and farmer-led innovation, adoption, and adaptation.

Kenya's capital city, Nairobi—sitting very near the equator at an altitude amenable to year-round fair weather—has played an important role in hosting international development agencies as well as international financial institutions, diplomatic and intergovernmental offices, military bases, non-governmental organisations, and private sector headquarters. These serve not only Kenya but also the larger eastern, central, southern, and Horn of Africa regions, which have been scarred by chronic war, famine, political instability, and impoverishment, including in Tanzania, Uganda, Somalia, Ethiopia, Eritrea, Sudan, South Sudan, Rwanda, Burundi, the Democratic Republic of the Congo, Zimbabwe, South Africa, and Mozambique. The combination of a rich diversity of enduring indigenous, as well as externally introduced, practices of collective action, on the one hand, and Kenya's geopolitical location amidst troubled states, on the other hand, has made Kenya a central hub for development and related research, planning, policy-making, and programming as well as knowledge production, mobilisation, and innovation. Important research advances in Kenya have helped cement the now almost universally accepted understanding of the relationship between food security and gender equity (Sanginga, Waters-Bayer, Kaaria, Njuki, & Wettasinha, 2009). Scholarship in the region has also contributed significantly to resilience thinking, in particular examining links between social and ecological resilience (Bahadur, Ibrahim, & Tanner, 2013; Folke et al., 2002; IFPRI, 2013).

In a part of the world that is no stranger to hunger and malnutrition, the focus for resilience work in East Africa has been importantly on agriculture and, more broadly, natural resource management (Alinovi, D'Errico, Mane, & Romano, 2010). But given the importance of women in agriculture and development, surprisingly little attention has been given, in East Africa and elsewhere, to the question of the connection between gender equity and socio-ecological resilience within farming systems. The topic has been partially addressed in the literature on women and land rights and sustainable development (e.g., Agarwal, 1994; Charkiewicz, Hausler, Wieringa, & Braidotti, 1994; Harcourt, 2012). This collection aims to bring the specific language of socio-ecological resilience to the analysis of Kenyan farmers' agency in the face of change. We see this as useful in offering road signs to effective means by which to achieve gender equity within local communities, not as an end in itself, but as a prerequisite for household food and nutrition security.

Despite its rich potentials, Kenya has not escaped the regional trend of high rates of hunger, malnutrition, and the negative health and economic consequences that follow. In Kenya's arid and semi-arid lands (ASAL), which account for over 7.5 million hectares and hold 20% of the country's population, declines in agricultural productivity in the recent past have been particularly striking (Ngugi & Nyariki, 2006). Low and erratic rainfall makes agriculture a challenge in the arid and semi-arid lands, which also face falling farm size and crop yields and uncontrolled conversion of forests and scrubland to agriculture, which has contributed to degradation of soil, tree cover, and water resources (Nyariki, Wiggins, & Imungi, 2002; Simpson, Okalebo, & Lubulwa, 1996). Kenya government statistics show that the proportion of individuals living on less than US$1/day in the ASAL regions stands at 65% (KNBS 2010). In Makueni County, one of the three counties included in our project, poverty levels are particularly high at 74% (KNBS, 2010). Hunger and malnutrition make their biggest impacts on women and children, with close to one-third of children suffering from malnutrition (KNBS, 2010).

Gender inequalities in land distribution and decision-making power are further barriers to the success of food security initiatives. Complex and diverse indigenous land tenure systems prevailed in Kenya prior to the colonial era (approx. 1890–1963). These tenure regimes ensured that virtually all members of the community had access and entitlements to some land and other livelihood resources, whether through recognised ownership, usufruct rights, temporary occupation, frontier expansion, tenancy, inheritance, marriage, adoption, or loan. These mechanisms provided several forms of entitlement for women, such as the provisions among the Kikuyu and other ethnic groups for unmarried single mothers to inherit land from their fathers (see Chapter 4, this volume; Mackenzie, 1998).

However, the rights of women and landless men were severely curtailed during the colonial era, specifically through the British administration's formalisation of Kenyan land tenure customs in state-sanctioned, written customary law. While many British colonies were not settler oriented, Kenya was viewed by London as an important white settler destination and promoted it as such, for instance in soldier settlement schemes after both world wars. The British organised several administrative studies and land commissions in the early 20th century, more or less all with the purpose of making "recommendations as to what rules should be enacted to govern the occupation rights of tribes, clans, families or individuals in each or any area, due regard being had to Native Law and Custom" (Colony and Protectorate of Kenya, 1929, p. 5).[2] These studies and reports provide important insight into both the range of land tenure norms practised in the early colonial era, as well as some of the blind spots that were to emerge in the reports' findings and in the eventual articulation of the practices in written customary law. Beech, for instance, examined the non-exclusivity of Kikuyu land tenure systems. He found that indigenous land relations embraced both individual and common rights to land in a system of overlapping entitlements. Furthermore, although he believed that no woman was able to inherit or own land, "there are elaborate rules as to a woman's life interest in lands, just as certain duties with regard to cultivation are her privileges and hers alone" (Beech, 1917, p. 56). The chiefs, headmen, and male elders who were interviewed in the various commissions emphasised in their testimonies those components of land tenure "that privileged male authority over the land. Thus, they initiated a dominant discourse within Kikuyu society, wherein the interests of women, and the landless, were unrepresented" (Mackenzie, 1998, p. 76).

Instead of accurately representing the array of indigenous land tenure arrangements that ensured near-universal access to land for both women and men, the regulations emerging from these reports limited private ownership of land to men, and mainly to the wealthier and more loyal chiefs, headmen and other elites. Kenya's history of exclusionary land legislation has not been very effectively reversed in the postcolonial era. In 2006, only some 5% of land titles in Kenya were held in the names of women (Deere & Doss, 2006). Fifty years after independence, millions of people whose grandparents were dispossessed in the colonial era continue to live insecurely in slum settlements in cities and rural areas. Postcolonial development policies further entrenched exclusionary practices. Of great importance here are World Bank structural adjustment programmes introduced in the 1980s that reemphasised the prioritisation of export cash-cropping by both large- and small-scale farmers. These policies contributed to the deepening of gendered inequalities in access to land and, in turn, to decisions over household livelihood strategies, in which most women farmers' prioritisation of food production could be trumped by male landowners' decisions to plant cash crops for export (Brownhill, 2007).

It is against this backdrop that the authors of this collection joined together in resilience-focused research in the semi-arid midlands of Eastern Kenya. Because of the complexity of the problem of household food security, we looked for answers not in single agricultural technologies but in the adoption of a range of integrated practices and technologies which address different requirements of the farming system, from inputs to outputs, from pre- to post-harvest. We included education on preparing nutritious foods, negotiating market access, and handling of crop residues and waste. In addition, our approach considered institutional and policy contexts, including household-level social dynamics, and farmer group organising as well as international, national, and county-level policy processes. By taking this integrated approach, we sought to understand the farming system's history and potential futures, in terms of science and policy's potential to support farmers' already existing knowledge and capacities to experiment with, adopt, and adapt resilience-building agro-economic practices and technologies.

Five seasons of farmer adoption, evaluation, learning, and adjustment of farm practices within the project's research sites[3] led to greater productivity and diversity of crops, improved incomes for women and men, better nutrition, and reduced periods of household food insufficiency. Another important measure of the success of the project lay in the improvement of women's control over resources (green grams and indigenous chicken, among others) and the food, nutritional, ecological, and income benefits derived from these enterprises (Njuguna, Brownhill, Kihoro, Muhammad, & Hickey, in press).

Socio-ecological resilience

Implementing such a diverse range of studies in one project was a challenge. Undertaking an analysis that integrated, and made sense of, the diverse resulting research findings was an even bigger challenge. With so many elements of the system to trace, the concept of socio-ecological resilience was a natural starting point for analysis, since it embraces and helps us keep track of, and explain, the farming system's complexity, as well as identify weaknesses and strengths in terms of building household food security.

In the 1970s, ecological resilience theory emerged from the fishponds and labs of ecologists (Holling, 1973) to be adopted and operationalised in almost every field of study. Resilience has rightfully become a key concept—and the strengthening of resilience a critical practical objective—in areas ranging from education, engineering, and psychology to harm-prevention counselling, disaster relief, and the building of enduring food security. Its core meaning is the capacity—of individuals, households, communities, cities, and whole societies—to recover from crises. This capacity takes on specific attributes, scales, and expressions within different fields. The wide applicability and development of the concept of resilience provides a wealth of disciplinary perspectives from which to enrich our own socio-ecological understanding of the term. The UN's Office for Disaster Risk Reduction defines resilience as the capacity of a system, community, or society, potentially exposed to hazards, to adapt by resisting or changing in order to reach and maintain an acceptable level of functioning and structure (UNISDR, 2009, p. 4).

The UN Framework Convention on Climate Change seeks to "reduce vulnerability and build resilience in developing countries," specifically in order to better face the onset of global climate change (UNFCCC, 2010, p. 4). Buzz Holling, the proclaimed father of resilience theory, maintained a philosophical air about the state of the planet's ecological and economic systems. He harked back to his own earlier description of "rare events," like hurricanes, that can "unpredictably shape structure" and potentially cause disruptions or system collapse. Holling used the term *resilience* to characterise the current era as one of "rare transformation," an era of "major economic, social and environmental transformation," in which resilience practices are ever-more urgently needed (Holling, cited in Ross, 2010, p. 8).

Resilience, as a concept and a process, provides the outline of an understanding to guide action to adapt to, manage, and shape such rare (uncommon, sudden, and systemic) transformations. Resilience contains within it recognition of the vulnerability of complex systems as they undergo (resist, adapt to) changes, of a periodic and incremental nature or in sudden and rare transformations. Holling's acknowledgement of the planetary transformation taking place in the 21st century serves to emphasise both the scale of ongoing socio-ecological crises and the urgency with which resilience thinking and resilience action are required, not only in communities known to be vulnerable to hunger and malnutrition, but also in those communities, such as coastal cities like Miami, in which vulnerability has largely been unknown (McKie, 2014).

Socio-ecological sciences have much to learn from how resilience has been applied in other disciplines. In the field of education, resilience is the ability of children to succeed academically despite risk factors that increase their likelihood of failure (Ungar, 2012). Themes of "resilient parenting" and "family resilience" arise here (DuMont, Ehrhard-Dietzel, & Kirkland, 2009; Walsh, 2012). Bryan identifies "protective factors" that can increase educational resilience in children, including "caring and supportive adult relationships" and "opportunities for meaningful student participation" in matters that affect them (Bryan, 2005, p. 220).

Similarly, counselling and peer education can build the *psychological resilience* of children living through civil conflict or domestic violence (Ungar, 2012). Interestingly, Ungar traces the emergence of an ecological perspective in studies of psychological resilience, which shifted the focus of theory and practice from the propensities and capacities of individuals to a consideration of the environment, or wider socio-ecological context, within which individuals act and exist. In his endorsement of Fernando and Ferrari's 2013 *Handbook of Resilience in Children of War*, Michael Ungar observes that "no longer can we explain resilience as something inside a child. It is, as shown in the many studies that are discussed with some of the most disadvantaged children in the world, something we create by making children's *social ecologies* safer and more nurturing" (Fernando & Ferrari, 2013, emphasis added).

It is also illuminating to identify what resilience is not. Folke et al. (2002) note that, "the antonym of resilience is often denoted vulnerability. Vulnerability refers to the propensity of social and ecological systems to suffer harm from exposure to external stresses and shocks.... The less resilient the system, the lower is the capacity of institutions and societies to adapt to and shape change" (Folke et al., 2002, p. 13). From psychology, we see that vulnerability "is part of the human condition, and with serious crises or persistent stresses we often can't simply bounce back or return to the old normal. Life may never be the same and we must construct a 'new normal' on our journey forward. Thus, *resilience involves struggling well*, effectively working through and learning from adversity, and attempting to integrate the experience into our individual and shared lives as we move ahead" (Walsh, 2012, p. 174, emphasis added).

Resilience as it is applied in these diverse fields of study can provoke new understanding and enrich our conceptual toolboxes for analysis within the field of food and agriculture. Where there is persistent risk of crop failure, for instance in arid and semi-arid lands in Africa, can the fostering of *protective factors* (such as peer-to-peer agricultural education or policy support for small-scale and subsistence agriculture) help farmers circumvent risk and increase their potential to succeed in difficult circumstances? Can the "family resilience framework" developed in counselling psychology lend to agriculturalists and development specialists its notions of "social ecologies," of resilience as "struggling well," or of constructing a "new normal" after social, economic, or climate changes?

Given this rich range of ways in which the term *resilience* is employed, it became a question of central importance in our work to define what *we* understood it to mean, in the context of the farming systems in which we worked. Because we focus on farming systems, we began by denoting our field of inquiry in terms of *gendered food security resilience*.

The study of resilience in relation to food security is rooted within understandings of the environmental sustainability of agricultural systems, and the capacity of soils and other ecological factors to recover after extreme weather events and other, human-induced shocks. From these beginnings emerged more socially focused resilience studies that turned the focus towards the social relations characterising the operation of the overall farming system (Alinovi et al., 2010).

For us, food security resilience implies *systemic functioning*, or the capacity of the farming system to persist in the face of change and to continue to serve its purpose of feeding people and providing livelihoods. At the same time food security resilience implies *elemental integrity*, or the efficacy of the distinct operations of the constituent elements of the system. With food security resilience, we place a strong emphasis on the social relations of small-scale and subsistence agricultural production, and small-farming people's interactions with and adaptation to prevailing ecological and social conditions.

We here turn to a consideration of the resilience characteristics that are included in our assessment and analyses. In addressing how elements of the farming system either contribute to or undermine the whole system's resilience, we rely on the work of Bahadur, Ibrahim, and Tanner (2010, 2013), who analysed the scientific literature to delineate 10 key characteristics of socio-ecological resilience (see Box 1.1).

With this resilience typology, we characterise the state, strengths, and weaknesses of the farming system under study, with reference to these 10 indicators. These characteristics emerge as consistently important in terms of their contributions to reducing social and ecological vulnerability in the face of climatic and economic shocks. In turn, strengthening these capacities contributes to building the enduring socio-ecological systems that are the intended results of food security efforts by farmers, researchers, and policy-makers in small-scale and subsistence farming systems.

Box 1.1 Characteristics of socio-ecological resilience.

1   Diversity
2   Effective governance, institutions, and policies
3   Acceptance of uncertainty and change
4   Community involvement and inclusion of local knowledge in planning and programs
5   Preparedness, planning, and readiness
6   High degree of equity
7   Shared social values, networks, and structures
8   Nonequilibrium system dynamics★
9   Learning
10   Adoption of a cross-scalar perspective

★the 'new normal'

Bahadur, Ibrahim, & Tanner, 2013

A gender-transformative approach

As mentioned above, while numerous scholars demonstrate the link between women's empowerment and greater food security (Deere & Doss, 2006; Quisimbing & McClafferty, 2006), less has been achieved in terms of conceptualising the link between gender equity and the resilience of farming systems. One notable exception can be seen in Singh and Faleiro (2013, p.14), who "recognise resilience as a potential opportunity to transform power relations, particularly for women smallholder farmers." Conversely, we also see gender equity as a key contributor to the building of resilience, and necessary to the achievement of wider systemic relations of equity, diversity, community involvement, and shared values.

It is our recognition of the *change factor* implicit in resilience thinking and action that leads us to a consideration of theoretical frames that embrace and encourage individual and group agency and capacity to manage and influence the direction of change. We adopt the "gender transformative approach" discussed by Njuki and Miller (2013) to examine power relations among women and men, young and old, in farming livelihoods at the sites of the implementation of the project. Transformed gender relations are here identified as those which *undo* the exploitative, hierarchical, and exclusionary social relations that are associated with top-down research and the persistence of the twin crises of hunger and malnutrition; and which *replace* these with the cooperative, equitable, and inclusive social relations more closely associated with participatory research and with greater success in efforts to establish food food-secure farming systems, practices, and technologies (Gurung, 2011; Kumar & Quisumbing, 2010; Njuguna et al., in press). A gender-transformative lens shifts research and development activities (as well as evaluation, market development, and policy focus) from *gender blind* and exploitative, to *gender transformative* and characterised by equity and empowerment (Figure 1.1).

Adopting this approach has immediate implications. By conceptualising transformation in this way, we enter the field of power and politics, in that we recognise

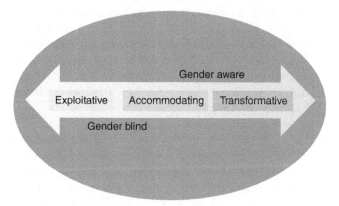

Figure 1.1 Gender-Transformative Approach.
After Njuki & Miller, 2013.

exploitation as emanating not only from asymmetrical power relations within the household, but also from the larger and even more powerful and entrenched power asymmetries within the community, nation, and global political economies in which farming households are set. As farmers, researchers, and policy-makers, among others, build resilience capacities that involve efforts to reduce power imbalances and make decision-making processes more horizontal, wider geopolitical power asymmetries can only be ignored at the peril of the intended resilience outcomes. When local trade networks, for instance, start to expand to the degree that they begin to cross paths with global market-oriented value chains, international trade agreements may end up putting a stop to such local market development, or reframing the developments more along global market lines.

This is a very real possibility, considering that the project studies and promotes a recognition of the strengths, expertise, and rich resources located within local farmers' indigenous knowledge, seeds, networks, crop varieties, farming practices, and ecological wealth. The socio-ecological resilience we describe and analyse in this collection is rooted in the elaboration of indigenous subsistence farming systems from below. It is a perspective distinctly opposed to the notion of linking subsistence farmers into global value chains at the expense of the development and servicing of very local markets and nutritional value chains (Hawkes & Ruel, 2012). The resilient farming system we discuss herein includes articulation with high-value traditional crops, very local markets, and agricultural value chains which retain labour, resources, products, and other values to first serve food security needs within the farm community, and then beyond the farm community in relatively short value chains. Thus the transformation we discuss here is not only about making women and men more equal within their households, but by so doing, making the entire political economy of farming more amenable to that equity, and to its implications for the development of very local value chains.

Transformation, like resilience, is not an end in itself. It is a means to an end. The end or goal of gender transformation is equity, as the goal of resilience is to replace those vulnerabilities, conditions, and relations which undermine the well-being of hunger-prone communities with conditions and relations that support and strengthen enduring well-being for all. Given that our project's objectives were concerned with improved household food and nutrition security, the transformation that we particularly sought was one which furthers the achievement of food security not for a few women and children in a few households but for the vast majority—that is, the general population.

Finally, while our research team was multidisciplinary and the design of the overall study was integrative, in practice many of the findings were published and presented in disciplinary journals, including in nutrition, economics, policy, and agronomy (see project website for a listing of publications, http://karimcgill-foodsecurity.org/). As senior gender researchers, the lead editors of this collection had a good position from which to examine cross-cutting socioeconomic concerns across the various research streams and disciplines. In our efforts to analyse how gender relations were impacted across project activities, we sought a unified framework within which to organise our understanding of the complexity of the farming system as well as the implications of project activities for gender relations within this system. As such, we needed to examine elements of the farming system through which gendered social relations are expressed, negotiated, and sometimes challenged and transformed. We identified four such sites which are, together, also descriptive of key elements of the farming system's organisation and operations. These sites are (i) socio-cultural relations of access and entitlements; (ii) availability of social, natural, and other resources; (iii) livelihood means, activities, and strategies; and (iv) institutional relations and policies which serve to support (or undermine) individuals' and groups' access to resources and livelihoods. We further consider these themes briefly in the section below, which describes this collection's contents, and then revisit them in more detail in the conclusion, where we assess their usefulness as organising concepts and guides to understanding and action.

Introducing the studies

The chapters in this collection present the perspectives of 21 contributors drawn from the fields of anthropology, agronomy, development, natural resource sciences, nutrition, plant science, law, and public policy. The collection of these studies into one volume serves two purposes. First, it creates a terrain in which results of studies from various perspectives are considered together, allowing lessons to be drawn from the insights afforded by our integration of the knowledge presented. It is not by merely appearing in the same edited collection, but by having focused and interacted for some years on a common research project (if from our distinctive disciplinary perspectives) that the contributors find a common ground. Secondly, the collection gave us the opportunity to develop and assess, with empirical data and case studies, an original methodological/conceptual framework which arose out of our collaborative efforts to understand and explain change in the farming system as a whole. This framework, the Resilience Umbrella, is detailed in the conclusion, where it is then employed in an analysis of the overall picture sketched out in these collected studies. Towards that end, we here categorise the studies according to their focus on the four key sites of gendered contestation within the farming system, as identified above: (i) access and entitlements, (ii) natural and social resources, (iii) livelihood activities, and (iv) policies and institutions.

In Chapter 2, Muhammad, Maina, Pelletier, and Hickey open the collection with a review of the participatory methodology through which the KALRO-KEMRI-McGill food security project was implemented, featuring on-farm trials of integrated crop, livestock, soil, and related agro-ecological practices. The farmer-led and group-based methodology, called Primary Participatory Agricultural Technology Evaluations, provided the basic research starting point, in relation to which each of the subsequent studies was designed and carried out in the field. This chapter reviews some of the project's overall impacts on gendered relations of access, resources, livelihood activities, and policies.

In Chapter 3, Eidt, Hickey, and Pelletier explore the relationship between farmers' practices of social networking and the success (or failure) of agricultural development research projects and programmes. By comparing two rural villages, one perceived as successful by farmers and local leaders in the areas, and the other perceived as unsuccessful and unwilling to participate in such projects, we show that stronger food security outcomes are associated with greater social networking, which serves to diversify men's and especially women's coping strategies and access to sources of agricultural knowledge relevant to their pursuit of food security objectives. The authors emphasise concerns about access to knowledge as a key component of farmers' capacities to learn, innovate, and adapt to change.

The subsequent three chapters focus on social and natural resources. In Chapter 4, Po and Bukania examine women's strategies to secure and maintain land resources as well as nutritional diversity in the face of economic and institutional constraints, and contention over unequal resource rights. The authors find that effective governance of land in semi-arid farming systems, for the food and nutrition security of vulnerable households (in this case, grandchildren who live with their grandparents), would include recognition and stronger formal legitimisation of various culturally defined forms of women's access to land, including usufruct rights, inheritance, and the custom of female-husbands. Protection of these entitlements would go some way towards giving women equitable access to direct sources of livelihoods. Though such a transformation entails wider socioeconomic changes, Kenya's new constitution provides an opportunity for county-level devolved decision-making, and thus an opening through which county residents can bring questions of women's land rights—and their connection to children's malnutrition—to ongoing civic and legislative processes.

In Chapter 5, Mucioki, Johns, and Mucioki focus on local farmer seeds varieties and the maintenance of what the authors term *matrilineal seed systems* in rural Tharaka-Nithi. Employing in-depth oral history as a key methodology, the authors catalogue Tharaka women's local pearl millet varieties, cultivation practices, harvest and post-harvest handling, as well as seed selection, saving, and sharing systems. Women are shown in this narrative to be effective stewards of biodiversity, indigenous knowledge, and of the cultural practices that underlay their entitlements to seed. The authors end by suggesting means by which matrilineal seed systems might be recognised and protected in law, and asking what role the local market might play in elaborating women's dissemination of locally adapted and highly nutritious food security crops such as pearl millet.

In Chapter 6, James and Po give attention to the financial resources that farmers draw on in pursuit of their livelihoods. The chapter explores both formal and informal micro-lending institutions, as well as the social capital instantiated in the predominantly informal institutions preferred by women farmers. They show that trust, participation, and inclusion are fostered in women's collective savings groups, and that these groups respond to women's particular credit needs, in terms of the relatively small loan amounts, low interest rates (if any), and group-based collateral. Policy and market efforts to improve women's access to agricultural credit are advised to embrace and support women's collective savings institutions as well as making women and farmer group-based loans more readily available in formal lending institutions and in women and youth-oriented government interest-free loan programmes.

The next chapter turns to a consideration of a key livelihood activity of small-scale and subsistence farmers in the semi-arid farming systems under study.

Chapter 7, by Brownhill, Njuguna, Mungube, Nzioka and Kihoro, emphasises the food security contributions of indigenous chicken enterprise for semi-arid farming households in Makueni. Backyard poultry rearing is shown to be an important livelihood activity for women. As a small-scale venture, poultry rearing brings immediate nutritional, economic, and ecological benefits and in this way appears nested within subsistence, market, and natural economies. The authors argue that the small-scale poultry enterprises studied here succeeded as a resilience-building measure precisely because of the farmers' subsistence orientation, with practices mutually supportive of health, income, and natural resources.

Next, two chapters highlight themes of policies and institutions.

In Chapter 8, Shumsky, Bothi, Nambiro, and Maundu address forestry policies that differentially impinge on subsistence livelihood activities of women and men. The chapter makes recommendations aimed at improving women's access to and stewardship of forestry resources for subsistence purposes that support biodiversity as well as household livelihood strategies.

In Chapter 9, Brownhill, Moturi, and Hickey consider the impact of policy changes pursuant to the devolution of governance, as introduced with the promulgation of a new constitution in Kenya in 2010. The chapter explores the potential for and challenges to the participation of farmers in agricultural policy-making, including a new agricultural policy presented for debate in 2015. The authors draw on their analysis of the findings of their mixed-methods study to make recommendations for the consideration of those policy makers who are involved in harmonising the various published guidelines for citizen participation in government decision-making. Key among the recommendations is the broadening of the diversity of participants who are encouraged, or even mandated, to engage in governance matters, based on constitutionally established norms and on the expressed views of farmers and other agricultural stakeholders interviewed here.

In Chapter 10, Brownhill and Njuguna discuss further findings of the INREF project through the lens of an original analytical tool, called the Resilience Umbrella, developed in this project for the purpose of analysing gendered dynamics across a complex farming system. This chapter first describes the Umbrella tool, then summaries some of the project's findings in light of the framing the Umbrella provides, and finally assesses the tool's usefulness. Does this framework offer something that is truly new? Does it tell us what can be done to strengthen on-farm practices and agriculture-related policies and services to more widely achieve enduring food security, gender equity, and socio-ecological resilience?

With this brief overview, we now turn to the substantive chapters covering the diverse topics mentioned above.

Endnotes

1 However, this collection does not exhaust the project's manifold findings, which are presented in more than 30 articles, conference papers, and reports that can be seen on the project's public website (http://karimcgill-foodsecurity.org/). The collection includes papers from project team members who responded to our closed Call for Papers and papers passed through blind peer review and editorial review processes.
2 These included a 1910 inquiry organized by early colonial administrator, and part-time ethnographer, Charles William Hobley (Hobley 1910); Mervyn Beech's 1912 study (Beech 1917); administrator John Ainsworth's 1920 land tenure report; the 1929 Maxwell Commission and the 1934–1935 Kenya Land Commission, also known as the Carter Commission.
3 The five seasons covered two and a half years.

References

Agarwal, B. (1994). *A field of One's Own: Gender and Land Rights in South Asia.* Cambridge: Cambridge University Press.

Alinovi, L., D'Errico, M., Mane, E., and Romano, D. (2010). Livelihoods strategies and household resilience to food insecurity: An empirical analysis to Kenya. Paper presented at the conference organized by the European Report of Development, Dakar, Senegal, June 28–30.

Beech, M. H. (1917). Kikuyu system of land tenure. *Journal of the Royal African Society,* 17(65), 46–59.

Bahadur, A. V., Ibrahim, M., and Tanner T. (2010). *The Resilience Renaissance? Unpacking of Resilience for Tackling Climate Change and Disasters.* (Strengthening Climate Resilience Discussion Paper 1). Brighton: University of Sussex, Institute of Development Studies.

Bahadur, A. V., Ibrahim, M., and Tanner, T. (2013). Characterising resilience: Unpacking the concept for tackling climate change and development. *Climate and Development,* 5(1), 55–65.

Brownhill, L. (2009). *Land, Food, Freedom: Struggles for the Gendered Commons in Kenya, 1870–2007.* Trenton: Africa World Press.

Brownhill, L. (Spring/Summer 2007). Gendered struggles for the commons: Food sovereignty, tree-planting and climate change. *Women and Environments International, Special Issue on Women and Global Climate Change, No. 74/75,* pp. 34–37.

Bryan, J. (2005). Fostering educational resilience and achievement in urban schools through school-family-community partnerships. *Professional School Counseling: Journal of the American School Counselor Association,* 8, 219–227.

Charkiewicz, E., Hausler, S., Wieringa, S., and Braidotti, R. (Eds.) (1994). *Women, the environment and sustainable development: Towards a theoretical synthesis.* London: Zed Books.

Colony and Protectorate of Kenya. (1929). *Native Land Tenure in Kikuyu Province* (Chairman G.V. Maxwell). Nairobi: Government Printer.

Deere, C. D. and Doss, C. (2006). *Gender and the Distribution of Wealth in Developing Countries.* (UNU WIDER—World Institute for Development Economics Research of the United Nations University, Research Paper No. 2006/115). Helsinki: UNU-WIDER.

DuMont, K., Ehrhard-Dietzel, S., and Kirkland, K. (2012). Averting child maltreatment: Individual, economic, social, and community resources that promote resilient parenting. In M. Ungar (Ed.), *The Social Ecology of Resilience: A Handbook of Theory and Practice* (pp. 199–218). New York, NY: Springer.

Elkins, C. (2005). *Imperial Reckoning: The Untold Story of Britain's Gulag in Kenya.* New York: Henry Holt and Company.

Fernando, C. and Ferrari, M. (Eds.). (2013). *Handbook of Resilience in Children of War.* New York, NY: Springer.

Folke, C., Carpenter, S., Elmqvist, T., Gunderson, L., Holling, C. S., Walker, B. et al. (2002). *Resilience and Sustainable Development: Building Adaptive Capacity in a World of Transformations.* (Scientific Background Paper on Resilience for the process of the World Summit on Sustainable Development on behalf of the Environmental Advisory Council to the Swedish Government). Stockholm: Environmental Advisory Council, Ministry of the Environment.

Government of Kenya. (2011). NALEP within the policy framework. *Biashara Leo,* April/May.

Gurung, B. (2011). Introduction: Engaging with the challenges for mainstreaming gender in agricultural research and development. In B. Gurung, E. Ssendiwala, and M. Waithaka (Eds.), *Influencing Change: Mainstreaming Gender Perspectives in Agricultural Research and Development in Eastern and Central Africa* (pp. 1–9). Entebbe: International Center for Tropical Agriculture (CIAT) and Association for Strengthening Agricultural Research in Eastern and Central Africa (ASARECA). Retrieved from http://www.asareca.org/resources/reports/booksandchapters/ASARECAGMIntroWEB.pdf

Harcourt, W. (Ed.). (2012). *Women Reclaiming Sustainable Livelihoods: Spaces Lost and Spaces Gained.* London: Palgrave Macmillan.

Hawkes, C. and Ruel, M.T. (2012). Value chains for nutrition. In S. Fan and R. Pandya-Lorch (Eds.), *Reshaping Agriculture for Nutrition and Health. An IFPRI 2020 book* (pp. 73–82). Washington, DC: IFPRI.

Hobley, C. W. (1910/1971). *Ethnology of A-Kamba and Other East African Tribes.* London: Frank Cass & Co.

Holling, C. S. (1973). Resilience and stability of ecological systems. *Annual Review of Ecology and Systematics, 4,* 1–23.

International Food Policy Research Institute (IFPRI). (2013). *Building Resilience for Food and Nutrition Security: Concept Note for an International Policy Cconsultation.* Washington, DC: International Food Policy Research Institute (IFPRI)/2020 Vision Initiative.

Kenya National Bureau of Statistics (KNBS) and ICF Macro. (2010). *Kenya Demographic and Health Survey 2008–2009.* Calverton, MD: KNBS and ICF Macro.

Kumar, N. and Quisumbing, A. (2010). *Does Social Capital Build Women's Assets? The Long-Term Impacts of Group-Based and Individual Dissemination of Agricultural Technology in Bangladesh.* (CAPRi Working Paper No. 97). Washington, DC: International Food Policy Research Institute. Retrieved from http://dx.doi.org/10.2499/CAPRiWP97

Mackenzie, A. F. D., (1998) *Land, Ecology and Resistance in Kenya, 1880–1952,* Portsmouth, NH: Heinemann.

McKie, R. (2014, July 11). Miami, the great world city, is drowning while the powers that be look away. *The Observer.*

Meinzen-Dick, R., Quisumbing, A., Behrman, J., Biermayr-Jenzano, P., Wilde, V., Noordeloos, M. et al. (2011). *Engendering Agricultural Research.* (IFPRI Monograph). Washington, DC: International Food Policy Research Institute.

Ngugi, R. K. and Nyariki, D. M. (2006). Rural livelihoods in the arid and semi-arid environments of Kenya: Sustainable alternatives and challenges. *Agriculture and Human Value, 22,* 65–71.

Njuguna, E.M., Brownhill, L., Kihoro, E., Muhammad, L. W., and Hickey, G. M. (in press). Gendered technology adoption and household food security in semi-arid Eastern Kenya. In J. Parkins, J. Njuki, and A. Kaler (Eds.), *Towards a Transformative Approach to Gender and Food Security in Low-Income Countries.* London, UK: Earthscan.

Njuki, J. and Miller, B. (2013). Making livestock research and development programs and policies more gender responsive. In Njuki, J. and P. C. Sanginga (Eds.), *Women, Livestock Ownership and Markets: Bridging the Gender Gap in Eastern and Southern Africa,* (pp. 111–128). London: Earthscan.

Nyariki, D. M., Wiggins, S., and Imungi, J. K. (2002). Levels and causes of household food and nutrition insecurity in dryland Kenya. *Ecology of Food and Nutrition, 41,* 155–176.

Presley, C.A., (1992). *Kikuyu Women, the Mau Mau Rebellion, and Social Change in Kenya*, Boulder, CO: Westview Press.

Quisumbing, A. R. and McClafferty, B. (2006). *Using Gender Research in Development*. Washington, DC: International Food Policy Research Institute.

Rathgeber, E. M. (1990). WID, WAD, GAD: Trends in research and practice. *Journal of Developing Areas, 24*, 489–502.

Ross, N. (2010). Calling Buzz Holling. *Alternatives Journal, 36*(2), 8–11.

Sanginga, P. C., Waters-Bayer, A., Kaaria, S., Njuki, J., and Wettasinha, C. (2009). *Innovation Africa: Enriching Farmers' Livelihoods*. London: Earthscan/Routledge.

Simpson, J. R., Okalebo, J. R., and Lubulwa, G. (1996). *The Problem of Maintaining Soil Fertility in Eastern Kenya: A Review of Relevant Research*. (ACIAR Monograph No. 41). Canberra: Australian Centre for International Agricultural Research.

Singh, H. and Faleiro, J. (2013). *Action Aid's Discussion Paper on Resilience*. Retrieved from http://www.actionaid.org/sites/files/actionaid/actionaids_discussion_paper_on_resilience_-_16_may_2013.pdf

Ungar, M. (Ed.). (2012). *The Social Ecology of Resilience: A Handbook of Theory and Practice*. New York, NY: Springer.

United Nations Framework Convention on Climate Change (UNFCCC). (2010). Report of the Conference of the Parties on its sixteenth session, held in Cancun from 29 November to 10 December 2010. Addendum Part Two: Action taken by the Conference of the Parties at its sixteenth session. Geneva, Switzerland, United Nations. http://unfccc.int/resource/docs/2010/cop16/eng/07a01.pdf#page=4

United Nations International Strategy for Disaster Reduction (UNISDR). (2009). *2009 Terminology on Disaster Risk Reduction*. Geneva: United Nations International Strategy for Disaster Reduction. Retrieved from http://www.unisdr.org/files/7817_UNISDRTerminologyEnglish.pdf

Walsh, F. (2012). Facilitating family resilience: Relational resources for positive youth development in conditions of adversity. In M. Ungar, (Ed.), *The social ecology of resilience: A handbook of theory and practice* (pp. 173–186). New York, NY: Springer.

Wipper, A. (1975). The Maendeleo Ya Wanawake organization: The co-optation of leadership. *African Studies Review, 18*, 99–120.

2 A participatory and integrated agricultural extension approach to enhancing farm resilience through innovation and gender equity

Lutta W. Muhammad, Immaculate N. Maina, Bernard Pelletier, and Gordon M. Hickey

Introduction

The smallholder farming communities operating in the semi-arid regions of Africa have been recognized as amongst the most vulnerable to climatic, environmental, and economic shocks and stressors internationally (De Souza et al., 2015). This is due, in part, to their reliance on pastoralism and rain-fed agriculture, which are very sensitive to climatic variability, continued environmental degradation associated with poor resource management, and growing populations (Cooper et al., 2008; Fraser et al., 2011; Sietz, Lüdeke, and Walther, 2011; Thornton, Van de Steeg, Notenbaert, and Herrero, 2009). In order to better address these issues, national and international development strategies are increasingly emphasizing the need to build social-ecological resilience and foster adaptive capacity in rural livelihoods (De Souza et al., 2015; Fan, Pandya-Lorch, and Yosef, 2014; Republic of Kenya, 2013).

As a key component of rural development, agricultural extension and advisory services (EAS) have the potential to play a significant role in strengthening the resilience of rural small-scale (often resource-poor) farmers against environmental- and climate-related shocks, and mitigate the consequences of market failure, poor governance, and related socio-economic vulnerabilities, by increasing access to resources such as knowledge and inputs (Davis, Babu, and Blom, 2014). EAS refers here to "all the different activities that provide the information and services needed and demanded by farmers and other actors in rural settings to assist them in developing their own technical, organizational, and management skills and practices so as to improve their livelihoods and well-being" (Sulaiman and Davis, 2012).

In Kenya, EAS delivery has been moving from the more traditional role of promoting the dissemination and adoption of agricultural technologies to that of strengthening innovation processes by facilitating social learning and collaboration among stakeholders while also building farmers' capacity to operate within the existing institutional environment (Chowdhury, Hambly Odame, and Leeuwis, 2014; Davis and Heemskerk, 2012; Klerkx, Van Mierlo, and Leeuwis, 2012; Sulaiman and Davis, 2012). Increased attention is also being given to gender-sensitive approaches to EAS delivery, as women and marginalized individuals have been shown to be frequently excluded from agricultural knowledge systems that lack adequate consideration of power relationships and access to resources (Farnworth and Colverson, 2015; Jafry and Sulaiman, 2013; Kingiri, 2013; Ragasa, Berhane, Tadesse, and Taffesse, 2013).

Although EAS are increasingly being conceived within the larger framework of agricultural innovation systems (AIS) (Davis and Heemskerk, 2012; Rivera and Sulaiman, 2009), the discourse linking social-ecological resilience, gender equity, and the provision of EAS is only emerging, and in need of greater empirical investigation (Davis et al., 2014). We view this need as part of the larger challenge of designing AIS that can both contribute to enhancing the adaptive capacity of smallholder farmers (Brooks and Loevinsohn, 2011; Leeuwis, Hall, van Weperen, and Preissing, 2013; Maina, Newsham, and Okoti, 2013) and address gender issues, particularly women's higher levels of vulnerability to shocks and stressors (Ifejika Speranza, 2006; Kumar and Quisumbing, 2014; Nelson and Stathers, 2009).

This chapter considers the role of EAS approaches to enhancing social-ecological resilience and gender equity in the rural communities of semi-arid Kenya by presenting a novel extension approach developed during a collaborative research project implemented by the Kenya Agricultural and Livestock Research Organization (KALRO), formerly the Kenya Agricultural Research Institute (KARI), and McGill University entitled *Innovating for Resilient Farming Systems in Semi-Arid Kenya* (INREF). The main objective of the project was to enhance food and nutrition security of rural households in semi-arid Kenya through the building of farming systems and institutions that could better support social-ecological resilience and gender equity.

In the sections that follow, we provide a brief overview of the EAS delivery systems used in Kenya, their underlying assumptions in terms of processes of agricultural innovation, and how this has informed the development of the INREF approach. We then pay particular attention to issues of gender and how they have been incorporated, or not, in the assessment of EAS effectiveness. The conceptual underpinnings of the INREF extension approach are then presented before we proceed with a description of project activities implemented in the field, and identify key project outcomes and policy implications.

Trends in the development of EAS models and their application

Many different models of EAS delivery have been implemented in sub-Saharan Africa and in Kenya in particular. The evolution of EAS delivery tends to reflect some important paradigm shifts that have taken place in agricultural research and development (Klerkx, Van Mierlo, et al., 2012; Pant and Hambly-Odame, 2009; Röling, 2009; Sanginga, Waters-Bayer, Kaaria, Njuki, and Wettasinha, 2009), while also occurring in response to difficulties achieving tangible impact 'on the ground' (Muyanga and Jayne, 2008).

An early dominant paradigm associated with agricultural extension is the *diffusion of innovation* proposed by Rogers (1962), which describes diffusion as a semi-autonomous process that can multiply the impact of research and extension within a community of farmers (Röling, 2009). Within this paradigm, innovations are first spread via communication within social networks of friends, relatives, and neighbours and then through the various messages prepared by extension personnel (Klerkx, Van Mierlo, et al., 2012). This view of agricultural innovation has led to the transfer of technology (ToT) model in which knowledge and technologies are generated by agricultural research centres and then disseminated to farmers via extension personnel in a generally linear, top-down fashion exemplified by the Training-and-Visit extension approach (Anderson, Feder, and Ganguly, 2006; Evenson and Mwabu, 2001). Furthermore, as highlighted by (Röling, 2009), the ToT model is also based on the view that diffusion of innovations can be triggered by market-led forces, as the increase in total production associated with the higher number of adopters would lead to a drop in price, which would then, in turn, force the rest of the farmers to either adopt or abandon.

The ToT model has been criticised for imposing standardized technical solutions that are not necessarily well adapted to the highly variable and risk-prone social-ecological environment within which African smallholder farmers operate, while also ignoring local knowledge and farmers' own capacity to innovate and experiment (Chambers and Jiggins, 1987; Rocheleau, 1999). The view that market forces could drive the innovation process has also been critiqued for not adequately taking into account the poorly developed institutions operating in many of these regions and their inability to support such innovation (e.g., input and output markets, micro-finance), thus potentially leading to increasing household vulnerability, especially among those with fewer assets (Röling, 2009). Despite these criticisms, however, it should be noted that the linear ToT model is still used as an important strategy in many agricultural research and development initiatives (Rivera and Sulaiman, 2009; Röling, 2009; Sanginga et al., 2009).

The relative failures of the ToT model in sub-Saharan Africa have made agricultural scientists more aware of the complexity and diversity of smallholder farming systems and the need to better involve farmers in the research and development process. The Farming Systems Research/Extension (FSR/E) movement, which emerged in the 1970s, proposed a more holistic perspective to study farming systems and factors affecting their performance, and a multidisciplinary approach to problem analysis (Cornwall, Guijt, and Welbourn, 1994). Approaches such as Farmer Participatory Research (Farrington and Martin, 1988) or Participatory Technology Development (Reijntjes, Haverkort, and Waters Bayer, 1992) were also proposed as processes of joint experimentation by researchers, development practitioners, and farmers that combine local and external knowledge to generate and disseminate agricultural innovations. Although participatory, these models are still typically driven by central authorities, such as governments, with support from international donors, and are usually grounded in policies, programs, and projects intended to spur rural development.

A clear departure from more centralized and linear EAS approaches has been the Farmer Field School (FFS) model widely described in the literature

(Braun, Jiggins, Roling, van den Berg, and Snijders, 2006; Sustainet EA, 2010). The FFS uses adult learning techniques and group-based learning through a scheduled programme from the beginning to the end of a crop-growing season. The approach has been identified as useful in helping farmers to deal with pests, to increase their incomes, and to gain knowledge (Najjar, Spaling, and Sinclair, 2013; van den Berg and Jiggins, 2007). Positive impacts of FFS on food security (Larsen and Lilleør, 2014) and the feeling of empowerment (Friis-Hansen and Duveskog, 2012) have also been reported. Despite these achievements, some studies (e.g., Davis et al., 2012) have indicated that FFS may have had limited impacts on economic performance and on farmer-to-farmer dissemination.

A more recent development is the focus on the potential role of EAS in AIS (Davis and Heemskerk, 2012; Klerkx, Schut, Leeuwis, and Kilelu, 2012; Rivera and Sulaiman, 2009). AIS can be defined as "a network of organisations, enterprises, and individuals focused on bringing new products, new processes, and new forms of organisation into economic use, together with the institutions and policies that affect the way different agents interact, share, access, exchange, and use knowledge" (Hall, Janssen, and Rajalahti, 2006). The AIS perspective does not consider research and technology development as the sole drivers of innovation, but instead views innovation as an emergent property of a learning, reflective, and transformative process involving multiple and interacting stakeholders and institutions (Hall et al., 2006).

Within the context of AIS, EAS thus need to go beyond knowledge brokering for research uptake and engage in *systemic facilitation* or *innovation brokering*, which includes a wide range of linkage building and facilitation activities aimed at transforming the technical, social and institutional relationships that drive innovation (Kilelu, Klerkx, Leeuwis, and Hall, 2011; Klerkx, Schut, et al., 2012). These may include tasks such as network building, supporting social learning, and dealing with dynamics of power and conflicts (Leeuwis and Aarts, 2011), in addition to helping farmers gain access to other resources needed for innovation—for example, capital, political support, business development services, and material resources (Klerkx, Schut, et al., 2012). Examples of innovation brokering functions are given in Kilelu et al. (2011), Rivera and Sulaiman (2009), Murray-Prior (2013), and Sulaiman and Davis (2012), among others.

In this type of facilitation role, EAS have the potential to contribute to the objective of building social-ecological resilience for food and nutrition security (Davis et al., 2014). The capacity of the various extension service providers (public, private, farmer-based) to implement such an approach represents, however, a major challenge. Numerous studies discuss the relative failures and inefficiencies of past and current extension models to accomplish these essential EAS functions and enable innovation (Mbo'o-Tchouawou and Colverson, 2014; Muyanga and Jayne, 2008; Poulton and Kanyinga, 2014). Challenges to scaling-up and -out this process need also to be systematically addressed, as many agricultural initiatives remain localized to limited project areas (Snapp and Heong, 2003).

Gender and agricultural extension

The availability, accessibility, and control over resources and assets (human, social, financial, physical, natural) needed to cope with shocks and build resilience (Alinovi, D'Errico, Mane, and Romano, 2010) are known to differ between men and women but also within and between households (Deere and Doss, 2006; Doss, 2001; Lim, Winter-Nelson, and Arends-Kuenning, 2007; Meinzen-Dick, Johnson, et al., 2011). Differences in assets among men and women may concern financial capital; ownership and user rights over resources such as land, water, livestock, grazing, and trees; the capacity to benefit from ecosystem services; labour use efficiency; and access to technology, training, and EAS (Ashby et al., 2012).

Access and control over these assets yield socioeconomic opportunities, which then underpin processes of empowerment—the condition in which individuals have the power to think and act freely, exercise choice, and fill their potential as fully equal members of society (Friis-Hansen and Duveskog, 2012; Mayoux, 2000). Women's empowerment is both a prerequisite and a desired outcome of AIS aimed at building resilience for food and nutrition security (Kingiri, 2013; Nelson and Stathers, 2009; Chapter 3, this volume). Because a complex set of factors—many of which are gender related—determines the way that male and female household members access and control assets, gender-sensitive approaches to resilience-building strategies are required (Ashby et al., 2012; IFAD, 2012; Kabeer and Natali, 2013; Kumar and Quisumbing, 2014; Nelson and Stathers, 2009).

In semi-arid Kenya, smallholdings fall into three major categories: male headed, female headed, and male headed but female managed, with a relatively high proportion of women-headed households. Even where the household is male headed, more often than not, much of the farm work is performed by the female spouse and other members of the household (DFID, 2007; World Bank, 2007). Despite this reality, many agricultural extension initiatives have failed to take into account appropriate gender considerations (Farnworth and Colverson, 2015; Jafry and Sulaiman, 2013; Meinzen-Dick, Quisumbing, et al., 2011; Ragasa et al., 2013; World Bank and IFPRI, 2010). In the more linear and top-down approaches, even in jointly managed farms, agricultural information received by husbands or other male household members is generally unlikely to reach women (World Bank and IFPRI, 2010). This gender bias is often observed when the content of extension service involves specific crops or enterprises in which commercial considerations are significant (Njuki, Kaaria, Chamunorwa, and Chiuri, 2011) or when tasks are divided by gender (Njuguna, Brownhill, Kihoro, Muhammad, and Hickey, in press).

A number of strategies to making EAS more gender-sensitive have been discussed (Chipeta, 2013; Jafry and Sulaiman, 2013; Mbo'o-Tchouawou and Colverson, 2014; Meinzen-Dick, Quisumbing, et al., 2011). They include (i) the strengthening of women's participation in organisations and leadership to ensure that EAS are in tune with women's needs and aspirations—this will also enhance the social and political capital of women and help address inequalities due to power relationships and rights to resources; (ii) recruiting and training women extension agents who can more easily reach other women; and (iii) targeting public administration by assigning positions to women—for example, gender focal points. Farnworth and Colverson (2015) advocate for EAS that are gender-transformative and able to empower women as agents of change alongside men in effective collaboration and decision-making processes regarding livelihood strategies and innovation. In this context, promoting gender equity becomes part of the 'innovation brokering' functions of EAS while potentially contributing to building social-ecological resilience (Bahadur, Ibrahim, and Tanner, 2013).

The INREF extension approach

As noted earlier in this chapter, the main goal of the INREF project was to enhance the sustainable food and nutrition security of rural households in semi-arid Kenya through innovative and resilient farming systems. Our proposed theory of change stated that this broader development goal would be achieved by (i) catalysing the adoption and scaling-up of farming practices to enhance food security and livelihood resilience; (ii) increasing household consumption of high-value traditional crops (HVTCs) that can improve the nutrition and health of children, women, and men; (iii) strengthening the participation of smallholder farmers into local and external input and output markets that allow women and men to diversify their livelihoods and improve their well-being; and (iv) contributing to the formulation of "resilience-focused" policies to enhance food security, livelihoods, and environmental sustainability while providing an enabling environment for innovation.

The project was based on the assumption that strong links exist between building resilient farming systems, gender equity, and the innovation potential of smallholder farmers. Consequently, project activities were conceived and implemented within the larger framework of creating an enabling and gender-sensitive environment for smallholder innovation based on the co-creation and sharing of knowledge among stakeholders through social learning activities, and the strengthening of smallholder social capital within farmer groups and through the building of linkages with key institutions—for example, input and output markets.

The project also adopted an integrated systems approach taking into account the multiple dimensions (environmental, social, economic, political, institutional) and scales influencing food and nutrition security in semi-arid regions (Graef et al., 2014; van Ginkel et al., 2013). Within this context, smallholder farmers and other stakeholders involved in the project participated in the generation and sharing of knowledge concerning the integrated management of agro-ecosystem components and functions (crops, soil, pests, water, livestock, biodiversity, labour) (Giller et al., 2011; Tittonell, 2014); agricultural product and household nutrition value chains (Hawkes and Ruel, 2012; Miruka, Okello, Kirigua, and Murithi, 2012); linkages between agriculture, nutrition, and health (Herforth, Lidder, and Gill, 2015); and institutional arrangements—for example, farmer groups, markets, and governance systems (Struik, Klerkx, van Huis, and Röling, 2014).

Participatory Agricultural Technology Evaluation

The basic organizational unit of the INREF approach is the Primary Participatory Agricultural Technology Evaluation (PPATE) cell. Membership of the PPATE comprises a group of about 15 to 50 households who live and carry out farming activities in the same locality and are well known to each other. The PPATE group evaluates technologies and practices that best fit its production, consumption, and market access systems while providing space for research, innovation, discovery, and joint learning with other stakeholders from among the research, extension, business, NGO, and smallholder farming communities.

The PPATE has affiliated to it one or more Secondary Participatory Agricultural Technology Evaluation (SPATE) groups (Figure 2.1), which also comprise approximately 15–50 households. Members of SPATE groups participate, observe, and learn from the various evaluation activities that are being implemented by the PPATEs and identify practices that they wish to try on their own farms. A major purpose of the SPATE component of the INREF approach is to facilitate assessments of impacts of proposed agricultural innovations on key social, economic, and environmental indicators and the resilience of rural livelihoods. This, we believe, is key to gauging the potential for adoption and identifying mechanisms for up-scaling of the innovations being evaluated. The new practices are then compared with the farmer's own crop or a neighbour's crop, grown on adjacent or nearby fields.

Implementation and evaluation of agricultural practices are guided by the Participatory Learning and Action Research (PLAR) conceptual framework (Defoer et al, 2009; Eksvärd and Björklund, 2010; WARDA, 2003), which is a "learning and innovation platform designed to bring together actors such as farmers, researchers and other stakeholders to jointly analyse farming and natural resource management issues, identify problems, seek, and develop solutions to those problems, and implement and evaluate these solutions, in an iterative learning-action cycle" (Figure 2.2).

In the PPATE/SPATE model, learning events are scheduled to be in alignment with crop development stages and address farming issues such as planting time, soil fertility and water management, and control of field pests and diseases. These events

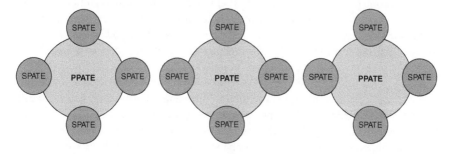

Figure 2.1 Schematic representation of an FRDA with three constituent PPATEs and the SPATEs.

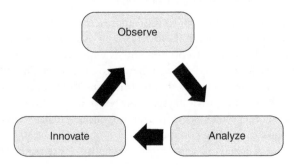

Figure 2.2 The INREF PPATE-SPATE PLAR cycle.
Defoer et al., 2009

become the main vehicle for engaging the participating farmer groups to plan, implement, analyse, inform, and receive feedback within the FRDAs, the PPATE and SPATE groups, and the general farming public. In contrast to many other participatory approaches to EAS, farmers are not considered as only 'recipients of new technologies' for them to adopt. Rather the intent is to create a process which would stimulate the farmers into discovering and innovating themselves. A series of farmer-to-farmer extension activities (Franzel, Sinja, and Simpson, 2014; Kiptot and Franzel, 2015) are also implemented in which 'champions' or 'service providers' are identified from among the PPATE/SPATE group membership to receive training on important components of rural livelihoods and value chains to share with their group or the larger community.

Implementation and outcomes of the INREF approach

Study area and site selection

The INREF approach was piloted in two agro-ecological zones (AEZs) in semi-arid Kenya, that is, AEZs Lower Midlands transition zone (LM4) and Lower Midlands semi-arid zone (LM5) in Machakos, Makueni, and Tharaka-Nithi counties (Jaetzold, Schmidt, Hornetz, and Shisanya, 2006). These areas are characterized by low and highly variable seasonal precipitation, and decreasing stocks of soil nutrients needed to sustain primary production and crop growth. In Kenya, the semi-arid AEZs account for about 20% of the human population (Jaetzold et al., 2006). In recent years, these regions have been experiencing environmental degradation and rapid population increases resulting in a reduction in average land holding size and farm productivity (Nyariki, Wiggins, and Imungi, 2002). Previous development efforts in the region have, however, identified numerous agricultural practices and technologies with the potential to sustainably enhance farm resource productivity.

Following stakeholder consultations in the three counties, sub-counties (former districts) exhibiting predominantly LM4 and LM5 AEZ conditions were selected and included: Makindu, Kathonzweni, and Makueni sub-counties in Makueni

County; Mwala and Yatta sub-counties in Machakos County; and Tharaka North and Tharaka South sub-counties in Tharaka-Nithi County. A total of 64 Focal Research and Development Areas (FRDAs) falling within AEZs LM4 and LM5 were then enumerated and designated as a "sampling frame" from which basic geographic units (FRDAs) for operation (Figure 2.3) were selected. Sub-county extension services convened open meetings called *barazas* involving Community Based Organizations (CBOs) representing geographical and sub-sector interests to select 18 FRDAs out of the initial 64. In each of the 18 FRDAs, CBOs conducted Focus Group Discussions (FGDs) to explain the INREF objectives and approach, clarify expectations, and identify three farmer groups that would each host a PPATE site. Thus, a total of 54 farmer groups to host PPATEs were recruited at these FRDA FGDs. Following another round of FGDs discussing the modalities of the INREF approach, all 54 PPATE groups identified one- to two-acre plots suitable for evaluating agricultural practices. In the majority of cases, the land was provided by a member of the PPATE group. In other cases, farmer groups secured the use of plots from public land following negotiations with local authorities or raised funds to lease the land.

Selection of agricultural innovations

In October 2011, FGDs were held in each of the 18 FRDAs to facilitate the scoring of 16 farming practices that could contribute to building resilience and gender equity (Table 2.1).

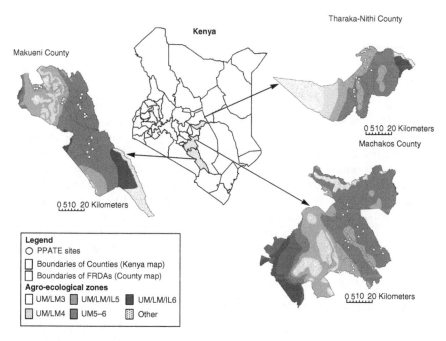

Figure 2.3 Map of the study area showing agro-ecological zones and location of PPATE sites.

Table 2.1 List of 16 farming practices to be scored and ranked in each FRDA in
          preparation of the PPATE evaluation activities

Type of crop or farming practice	*List of crops or farming practices*
Crops that had been introduced comparatively recently (4)	1. Maize (*Zea mays* L.) 2. Dolichos lablab (*Lablab purpureus* (L.) Sweet) 3. Common beans (*Phaseolus vulgaris* L.) 4. Grain amaranth (*Amaranthus spp.*)
Root crops (2)	5. Cassava (*Manihot esculenta* Crantz) 6. Sweet potatoes (*Ipomoea batatas* (L.) Lam.)
Traditional cereal crops (2)	7. Sorghum (*Sorghum bicolor* (L.) Moench) 8. Pearl millet (*Pennisetum glaucum* (L.) R. Br.)
Traditional grain legumes (3)	9. Green grams (*Vigna radiata* (L.) R. Wilczek) 10. Pigeon peas (*Cajanus cajan* (L.) Millsp.) 11. Cowpeas (*Vigna unguiculata* (L.) Walp.)
African leafy vegetables (1)	12. Vegetable amaranth (*Amaranthus* spp.); African black nightshade (*Solanum nigrum* L.)
Medicinal crop (1)	13. *Aloe vera* ((L.) Burm. f.)
Non-crop practices (3)	14. Improved management of indigenous chickens (*Gallus gallus domesticus*) 15. Natural pasture improvement 16. Fodder bank with napier grass (*Pennisetum purpureum* (L.) Schumach)

Each practice was then scored against four criteria: contribution to household *equity*; contribution to household cash *income*; contribution to *food sufficiency* (production, nutrition); and contribution to *resilience* of the farming system. For each FRDA, the scores obtained from each individual participant were averaged across the four criteria and then used to rank and identify the eight to ten practices that best suited the collective group preference within an FRDA. Results were disaggregated by gender to consider and discuss potential differences. The technologies or practices selected by each of the 18 FRDAs are presented in Table 2.2. It should be noted that in most FRDAs the practices preferred by women and men were often very similar, if not the same, even though the ranking may have differed. Each of the selected practices was then implemented following management and production guidelines and evaluated throughout the duration of the project. Details are provided in the following sections.

Managing cropping systems for building resilience

Local practices relying on low-quality seeds that are ill-suited for the areas where they are grown are known to contribute to food insecurity. Furthermore, largely because of the low use of complementary inputs such as fertilizers, pest and disease

Table 2.2 Participatory selection of agricultural innovations for PPATE evaluation

FRDAs	Coupeas	Ind. Chickens	Green Grams	Maize	Pigeon peas	Beans	Dolichos	Sweet Potatoes	ALVs[a]	NaPIs[b]	Cassava	Sorghum	Grain amaranth	Pearl Millet	Fodder/forages	Watermelon[c]	Aloe
Makueni County																	
Makueni sub-county																	
Watema	■	■	■	■		■		■		■	■				■		
Kivani		■	■	■	■	■		■	■	■					■		
Wote	■	■	■	■	■	■		■			■				■		
Kathonzweni sub-county																	
Thavu	■	■	■	■	■	■	■			■		■					
Mavindini	■	■	■	■	■	■	■			■					■		
Kithuki	■	■	■	■	■	■	■		■	■							
Makindu sub-county																	
Makindu	■	■	■	■	■	■	■		■			■	■				
Nguumo	■	■	■	■	■	■	■		■			■	■				
Machakos County																	
Mwala sub-county																	
Miu	■	■	■	■	■	■		■	■		■						
Kyawango	■	■	■	■	■	■		■			■						
Kavumbu	■	■	■	■		■		■			■	■					
Yatta sub-county																	
Kinyaata	■	■	■	■	■	■		■			■					■	
Ndalani	■	■	■	■	■	■				■	■	■					
Katangi	■	■	■	■	■	■	■	■			■						

(*continued*)

Table 2.2 Participatory selection of agricultural innovations for PPATE evaluation. (*continued*)

FRDAs	Coupeas	Ind. Chickens	Green Grams	Maize	Pigeon peas	Beans	Dolichos	Sweet Potatoes	ALVs[a]	NaPIs[b]	Cassava	Sorghum	Grain amaranth	Pearl Millet	Fodder/forages	Watermelon[c]	Aloe
Tharaka-Nithi County																	
Tharaka North sub-county																	
Thiti	■	■	■	■	■	■	■					■		■			
Tharaka South sub-county																	
Ntugi	■	■	■	■	■				■	■		■		■			
Nkarini	■	■	■	■	■	■						■		■			
Nkondi	■	■	■	■	■			■	■			■		■			
TOTAL	17	18	18	18	16	16	7	9	5	9	9	8	2	4	4	1	0

[a] African leafy vegetables; [b] Natural pasture improvement; [c] FRDA members added it to the list of 16 technologies.

control practices, and water conservation, farmers who do adopt high-yielding varieties often fail to realize the benefits from investing in certified seeds of such varieties (Ogada, Muchai, Mwabu, and Mathenge, 2014).

For each of the crops selected, the evaluation involved the comparison of two recommended improved varieties plus a local variety. The set of agronomic practices for the three varieties were the same and included integrated soil fertility and water management (ISFWM), and integrated pest and disease management practices (IPM), both considered key components of climate-resilient agricultural strategies (Campbell and Thornton, 2014; Lipper et al., 2014)

In terms of ISFWM, best practices to efficiently apply inorganic fertilizers and animal manure were discussed with PPATE participating farmers. In the second season (long rains of 2012), a number of on-farm plots were added to compare the efficiency of different water harvesting techniques—tied ridges, open contour ridges, and flat land. Following the joint assessment of agronomic performance due to the ISFWM, three practices proved to be most popular among farmers: the use of *boma* (farmyard) manure, efficient use of inorganic fertilizers, and in-situ rainwater harvesting. This result suggests a departure from standard practice in the area, as many farmers, especially in Tharaka-Nithi County, considered that *boma* manure should not be used because it encouraged growth of noxious weeds, and that the application of inorganic fertilizers could have negative effects on soils. The comparative study of water-harvesting practices suggested the potential of tied-ridges to increase crop yield significantly (Gichangi et al., 2007).

The losses that smallholder farmers incur due to crop field infestation by pests and diseases have been well documented (Abate, Van Huis, and Ampofo, 2000; Grisley, 1997) and are projected to worsen with climate change (Smith, 2015). A series of FGDs in all FRDAs were implemented to assess the pest and disease challenges faced by smallholder farmers and the coping strategies being implemented. Outcomes of these FGDs informed the design of on-farm demonstration workshops implemented by PPATE and SPATE groups throughout the growing season, which led to many farmers better appreciating and embracing IPM practices within the relatively short period of four rain seasons (two years).

Natural pasture improvement and fodder banks for livestock

In view of the importance of livestock in the mixed-cropping systems of semi-arid regions and the potential impact of climate change on feed crops and grazing systems (Thornton et al., 2009), the project included a NaPI (natural pasture improvement) component in which improved grass mixtures were established as a means to increase the productivity and quality of pastureland for both livestock grazing and land rehabilitation (Mganga, Musimba, Nyariki, Nyangito, and Mwang'ombe, 2015). In many of the PPATEs where these NaPI were established, however, results were relatively poor, leading some farmer groups to abandon the practice after a few seasons, indicating the need to further assess their performance. In terms of fodder bank establishment with napier grass, the recommended practices involved variety selection, the use of *tumbukiza* pits, and application of *boma* manure. PPATE farmers in the Makueni sub-county observed that the yield of napier grass was, on average, 48% higher in the *tumbikiza* pit than on flat land.

Improved management of indigenous chickens

Indigenous chickens (IC) are an important component of rural livelihoods in sub-Saharan Africa (Guèye, 2000) and in Kenya (Magothe, Okeno, Muhuyi, and Kahi, 2012; Okeno, Kahi, and Peters, 2012) contributing to income, wealth, dietary diversity, cultural and social activities, and insurance against shocks by providing food and/or income throughout the year. Because rural family poultry is predominantly managed by women, efforts to enhance IC management have the potential to contribute to gender equity (Guèye, 2000; Chapter 7, this volume).

Free range is the main production system found in rural Kenya (Okeno et al., 2012) and usually involves the birds scavenging on their own around the homestead. It is also characterized by low productivity and birds' vulnerabilities to parasites and predators (Mungube et al., 2008) and diseases such as Newcastle (NCD) (Magothe et al., 2012). A number of development initiatives have subsequently been implemented to address the main production and marketing constraints of rearing IC (Mailu et al., 2012).

The main IC management practices evaluated in the INREF project were recommendations for breeding, housing, feeding, and health care. Various approaches to collective (group) selling were also assessed. Each of the 54 PPATEs received 10 hens and a cockerel from the Poultry Research Unit based at KALRO Naivasha and used this initial batch of birds not only for evaluation and learning purposes, but also as a base for upgrading the local IC bird population. Each group nominated an IC service provider (ICSP) to participate in capacity-building events on its behalf. Representatives of women's groups from the Makueni County also joined the training for a total of 61 ICSPs (24 men, 37 women). The resulting pool of 61 ICSPs subsequently went beyond servicing their PPATE and SPATE members to include their immediate communities, with some agreeing to pay fees in order to make the process sustainable.

Agroforestry

A number of PPATE farmers from Tharaka-Nithi County expressed their interest in implementing agroforestry practices. The potential role of agroforestry in building resilient farming systems via the procurement of various ecosystem services is widely recognized (Mbow et al., 2014). In September 2012, following a series of consultations with various PPATE groups, capacity-building events were implemented by 13 PPATE farmer groups and included information on the sourcing of propagation material and technical aspects associated with tree husbandry. Over the course of the project, a total of about 12 tree species were included in the nurseries, providing a wide range of products and services (fruit, firewood, medicinal, fodder, shade, soil conservation). The agroforestry component of the project was primarily led by women farmers, giving them an opportunity to generate cash income, contribute to household food and nutrition security, access valuable products and services, and build social, technical, and political capital. In view of the challenges typically faced by women in implementing agroforestry initiatives (Kiptot and Franzel, 2012), the INREF project empowered them to participate and exercise leadership in a potentially key component of building resilience in semi-arid areas.

Post-harvest storage

Post-harvest grain loss of cereals and legumes is considered a major constraint to enhancing food and nutrition security in sub-Saharan Africa (Zorya et al., 2011) and is expected to be further exacerbated by climate change (Stathers, Lamboll, and Mvumi, 2013). Important health issues are also associated with excess moisture in stored grains, which worsens the effects of aflatoxin (a potent fungus toxin that has led to contamination, human illness, and, in some cases, fatalities in Eastern Kenya) if present (Daniel et al., 2011). To address these issues, we adopted a farmer-to-farmer extension model in which members of the INREF team (farmers and research assistants) and other extension personnel were trained in identifying storage pests and diseases, and the practices required to address the situations. Farmer representation came from all the groups in the three counties (62 men, 32 women). Following these workshops, field research assistants in liaison with farmers representing the FRDAs set up demonstrations in the different FRDAs, which reached an additional 234 farmers (113 men, 121 women) by the end of the project.

Access to quality seeds

One of the main constraints to the increased adoption of improved crop varieties by smallholder farmers in Eastern Kenya is the lack of access to certified seeds because of prohibitive cost, distance to seed providers, and/or lack of production of desired varieties (Muhammad et al., 2003). This issue concerns many HVTCs, which tend to fall outside formal seed systems (Muthoni and Nyamongo, 2008; Nagarajan, Audi, Jones, and Smale, 2007), highlighting the need to also better understand the contribution of informal seed systems (Muthoni and Nyamongo, 2008; Chapter 5, this volume).

To address this issue, nine farmer groups received training and initiated work with Freshco Seeds, a private-sector producer and distributor of certified seeds in Kenya, and the KALRO Seed Unit as part of a pilot project to establish commercial, community-based seed systems in the region. These initiatives were successful in creating capacity among group members in managing farm-produced seeds and selling them to seed companies. An increasing demand from the general farming community for the certified seeds promoted in the project suggests that the project's impact reached beyond participating farmer groups.

Access to produce markets

Enhanced access to produce markets by smallholder farmers is also considered an important component of diversifying household livelihoods and increasing the resilience of farming communities in semi-arid Kenya (Shiferaw et al., 2014) by stimulating adoption of improved practices, facilitating access to inputs, generating income, and boosting household financial resources. Yet market access by rural households in the semi-arid regions of Kenya is plagued by a lack of information on timing, quantities available and prices, and inadequate infrastructure.

To address these issues, INREF included a Participatory Market System Development (PMSD) (Griffith and Osorio, 2008) component to facilitate effective market access at the grass-roots level. In August and September 2012, farmers from the 54 PPATE groups, research assistants, and extension personnel participated in a workshop on PMSD promoting skills and knowledge on identification and assessment of market opportunities, and the assembly and bulking of produce for collective selling. These learning events were facilitated by KALRO, the local Ministry of Agriculture extension services and Cascade Development Organization, a Kenyan NGO promoting the development of inclusive markets in rural areas. In total, 716 men and 1,007 women received the training.

Members of the three PPATEs (and affiliated SPATEs) within each of the 18 FRDAs were then brought together as Marketing Opportunity Groups (MOGs), which then identified three priority commodities for their respective FRDAs—the majority selected green grams, indigenous chicken, and to a lesser degree, cowpeas. For each enterprise, MOGs conducted a market assessment (demand for produce, prices) and then drew up plans on the acreages that individual members were to plant, expected yields, and the required agronomic practices. The implementation of these production and marketing plans over three seasons resulted in new market-oriented behaviours across our study sites, which also led to higher produce prices for farmers and increased local household access to green grams and IC.

Working through their respective MOGs, farmer groups were able to identify and engage with numerous players in the farm produce market, including two firms, Smart Logistics Ltd. and Amerti Enterprise Ltd., which saw the value of the INREF approach and responded by entering into arrangements with the MOGs. Amerti worked with MOGs in the first season (October 2013–February 2014) aggregating produce, mainly green grams, for export to external markets in India. In the second season (March–August 2014), Amerti implemented a Collaborative Contract Farming System under which participating smallholders undertook to keep a "crop diary" to chart their acreage, seed type, and fertilizer use, and report on crop progress and previous seasons. Farmers would source knowledge and information on crop husbandry, farming techniques, and the use and source of proper seeds from KALRO and its partners. Amerti would provide skills and information on particular challenges facing the Kenyan agro-businesses in regard to the quality and quantities of farm produce that could address demand by the market, the need for traceability, and the market price fluctuations and seasonal demands. Amerti would be responsible for arranging financing to off-load the farmers produce. This arrangement allowed over 3,000 farmers to access markets for their produce on a sustainable basis.

Food and nutrition security

It is recognized that the building of resilience in rural livelihoods requires the improvement of linkages between agriculture, nutrition, and health (Fan, Pandya-Lorch, and Fritschel, 2012; Herforth et al., 2015). Nutrition is indeed both a contributor to, and a result of, enhanced resilience, as healthier individuals will generally be better able to cope with shocks and stressors (Dufour, Kauffmann, and Marsland, 2014). Pathways linking agriculture to better nutrition are usually described in terms of the enhanced capacity to purchase food following increases in farm income and/or through the increased production of more nutritious food for personal consumption (Haddad, 2013). These pathways are also strongly influenced by gender dynamics (Haddad, 2013; Meinzen-Dick, Behrman, Menon, and Quisumbing, 2012), as many studies have highlighted that women tend to spend more of their income on food, health care, and education of their children and that agricultural interventions that improve women's access to and control of assets are more likely to also enhance household nutrition (Meinzen-Dick et al., 2012). Both pathways were included in the INREF project via the identification of market-oriented value chains aimed at local and external markets and nutrition-sensitive value chains (Hawkes and Ruel, 2012) built around the increased household consumption of the HVTCs and IC products and the implementation of an awareness campaign on nutrition, health, and value-addition.

As part of the INREF project, the Kenya Medical Research Institute (KEMRI) developed a course comprising basic education on the importance of nutrition, the identification of various food types, and how to create a balanced diet. Designed as a training-of-trainers course, it involved a total of 121 (80 women, 41 men) nominated *Farmer Nutrition Champions* (FNCs) from the counties of Machakos and Makueni in February and March 2014. Follow-up evaluations of 15 FNCs in April 2014 showed that within two months of training, the FNCs had held 53 training sessions, reaching 715 community members with nutritional and dietary information.

In a household survey (n = 405) performed toward the end of the project, PPATE farmers reported an improvement of their food security situation, expressed as the number of months with sufficient food before and after the project, compared to non-participating farmers from the same sub-counties (Njuguna et al., in press). The extent to which this finding can be generalized to all project participants requires further research and evaluation.

Farmer field days

Another important component of the knowledge management strategy was the implementation of a series of Farmers' Field Days (FFDs), a common EAS tool among organizations involved in extension activities (Amudavi et al., 2009; Franzel et al., 2014). In total, 24 FFDs were organized in the three counties and brought together just under 6,000 participants from farming communities, local authorities, and professionals from local research, extension, and business communities (Table 2.3). Each FFD was organized around the visit of one of the PPATE sites and provided an opportunity for the larger farming community to observe, learn, and share knowledge about the farming practices being evaluated. The active participation of multiple stakeholders also contributed to the exchange of knowledge on a wide range of issues of relevance to smallholder farmers.

Pilot initiative on scaling-up

One stated objective of the project was to address the challenges associated with the scaling-up and -out of agricultural initiatives. Although the project had been quite successful among participating farmers, questions remained regarding the resources required to consolidate the accomplishments with participating farmers but also reach new farmers outside project areas. The challenges of scaling-up highly participatory development initiatives have been previously highlighted (Snapp and Heong, 2003).

Table 2.3 Total number of Farmer Field Days and male and female participants in the three counties

Year	Number of events	Number participating		
		Male	Female	Total
2012	13	1334	1229	2563
2013	4	441	790	1231
2014	7	610	1244	1854
Total	24	2385	3263	5648

One of the most significant outcomes of the project is that some key players in the smallholder sector from all seven sub-counties hosting PPATE sites took steps to replicate the INREF approach in new areas. Convinced of the value of the approach to disseminate agricultural practices, they initiated a *scaling-up pilot project*. In the October 2013 to March 2014 season, extension personnel from the Ministry of Agriculture (MoA) from the seven sub-counties identified scaling-up FRDAs with no previous experience with the INREF methods and replicated some of the procedures used in the PPATE and SPATE identification process and also for farmer evaluation of recommended agricultural practices. Two to three farm practices that were assessed as having shown the greatest potential to contribute to marketability, household incomes, and food security were identified from the INREF PPATEs and used in the scaling-up pilot project. Radio stations in both the Ukambani and Tharaka-Nithi sub-counties were identified to disseminate information on improved farm practices. In total, 2102 household representatives (715 women and 1387 men) in 61 groups had exposure to resilience-enhancing agricultural practices through implementation of the INREF approach. Over a relatively short two seasons, the number of female and male farmers directly participating increased by 78% and 66%, respectively (Table 2.4).

Discussion and conclusions

In examining the main features of past EAS efforts, it emerges that innovation is created through the interplay of ideas from multiple sources across different scales, from grass-roots initiatives to global programming. To better tap into these sources, farmers need to gain the necessary individual and organizational capacities to collaborate with other players in developing approaches that are appropriate to their situation (Klerkx, Schut, et al., 2012). This growing acknowledgement of the multifaceted nature of innovation systems has led to recent research and policy efforts to stimulate synergy between various partners (through projects and organizations) in agriculture research, extension, and education to better transform their ideas and products into innovations for the benefit of many.

Table 2.4 Number of farmer groups and male and female group members participating in the INREF pilot project on scaling-up.

County	Sub-County	Number of groups		Number of members					
				Female		*Male*		*Total*	
		Season		*Season*		*Season*		*Season*	
		1	2	1	2	1	2	1	2
Machakos	Mwala	4	13	41	178	106	409	147	587
	Yatta	–	6	–	46	–	90	–	136
Makueni	Kathonzweni	3	3	28	28	72	72	100	100
	Makindu	20	23	242	295	465	529	707	824
	Makueni	9	9	90	90	193	193	283	283
Tharaka	Tharaka North	–	3	–	29	–	56	–	85
	Tharaka South	–	4	–	49	–	38	–	87
	Total	36	61	401	715	836	1387	1237	2102
	% increase		69		78		66		70

The INREF extension approach was designed to facilitate a more gender-sensitive process of joint learning and co-creation of knowledge among smallholder farmers and other stakeholders with a view to identifying and implementing technical, social, and institutional innovations in support of household and community resilience (Klerkx, Schut, et al., 2012; Kristjanson, Harvey, Van Epp, and Thornton, 2014). This approach was based on a highly participatory process in which both men and women's perspectives, constraints, and aspirations were explicitly included in the collective design and implementation of project activities (Farnworth and Colverson, 2015).

The PPATE/SPATE model of the INREF approach provided a knowledge and information support function for the participating farmer groups. The model encompassed a system for the transfer and exchange of practical information on crop and livestock production as well as natural resource management by diverse rural development actors in semi-arid Kenya. Key to the approach was ensuring that the proposed resilience-enhancing agricultural practices were not *pushed* for farmers to *adopt* but used as tools to stimulate discussion, learning, and innovation among participating farmers. This was instrumental in facilitating an appreciation of, and interest in, the farm practices that had otherwise been promoted through agricultural extension services for many decades with variable success. The creation of a cohort of *farmer-trainers* from within the participating groups on nutrition and health, IC, PMSD, post-harvest storage, and agroforestry also contributed to strengthening farmer-to-farmer dissemination and sharing of knowledge.

Each of these extension functions sought to promote knowledge, change attitudes, improve skills, and expand aspirations (Rivera and Qamar, 2003). The high level of participation we observed in implementation indicates that the approach was successful in creating genuine interest and engagement among women and men farmers. In particular, women were generally enthusiastic participants in PPATE/SPATE activities in terms of their numbers, their active involvement in selecting and evaluating innovations, and taking leadership on a number of initiatives (e.g., IC, agroforestry). In our view, this contributed to a better integration of issues related to nutrition-sensitive value chains (Haddad, 2013; Meinzen-Dick et al., 2012), and enabled gender-related issues to be explicitly considered in the identification and design of resilience-focused initiatives (Nelson and Stathers, 2009).

The social learning activities built in to the INREF approach were not only a way to share knowledge among stakeholders but also contributed to building farmers' social capital through the creation or strengthening of relationships between farmers and other stakeholders, and across institutions (Bernier and Meinzen-Dick, 2014). The PPATE/SPATE model therefore facilitated a degree of political, social, and economic organization through which community groups co-managed their crop and livestock production and marketing activities. Ultimately, farmer groups were strengthened, an outcome that has already been identified as one of the *most effective pathways of innovation diffusion* in semi-arid Kenya (Darr and Pretzsch, 2008). Furthermore, the creation and strengthening of MOGs at the FRDA level increased the number of farmers accessing external and local markets, highlighting the important role of collective action for enhancing smallholder farmers' market access (Fischer and Qaim, 2012; Markelova, Meinzen-Dick, Hellin, and Dohrn, 2009; Shiferaw, Hellin, and Muricho, 2011) and possible mechanisms for engaging the private sector (Ferroni and Castle, 2011; Poulton and Macartney, 2012).

By embracing an integrated systems perspective from the start, the INREF approach enabled EAS learning and capacity-building activities on a wide range of issues related to sustainable agro-ecosystems and agricultural product and nutrition-sensitive value chains with a view to identifying potential bottlenecks and trade-offs. This also contributed to the identification of a number of *gendered* issues that needed to be more thoroughly addressed in the context of building resilience, such as the management of livestock and pasturelands (Thornton et al., 2009); post-harvest storage (Zorya et al., 2011); agro-biodiversity (Johns, Powell, Maundu, and Eyzaguirre, 2013); management of forest resources (Shumsky, Hickey, Pelletier, and Johns, 2014; Chapter 8, this volume); micro-finance (Chapter 4, this volume); and land (Jayne, Chamberlin, and Headey, 2014; Chapter 6, this volume). It should be noted that although the need for integrated and systems-thinking approaches has been well discussed in the broader context of agricultural research and development (Ericksen, 2008; Foran et al., 2014; Graef et al., 2014; Misselhorn et al., 2012; Uphoff, 2014; van Ginkel et al., 2013), its explicit incorporation into EAS systems has not, to our knowledge, been adequately addressed.

Despite the many difficulties associated with assessing the direct impact of EAS and social learning activities, (Aker, 2011; Davis, 2008; Kristjanson et al., 2014; Larsen and Lilleør, 2014), the INREF extension approach presented in this chapter offers an important opportunity to reflect on the potential role of EAS in facilitating innovation is support of gender-sensitive resilience in the food-insecure smallholder farming contexts of semi-arid Kenya. The field results also support the need for EAS to better integrate gender considerations into participatory and social learning activities designed to build social capital and networks in support of agricultural innovation.

References

Abate, T., Van Huis, A., and Ampofo, J. (2000). Pest management strategies in traditional agriculture: An African perspective. *Annual Review of Entomology*, *45*(1), 631–659.

Aker, J. C. (2011). Dial "A" for agriculture: A review of information and communication technologies for agricultural extension in developing countries. *Agricultural Economics*, *42*(6), 631–647.

Alinovi, L., D'Errico, M., Mane, E., and Romano, D. (2010). Livelihoods strategies and household resilience to food insecurity: An empirical analysis to Kenya. Paper presented at the conference organized by the European Report of Development, Dakar, Senegal, June 28–30.

Amudavi, D. M., Khan, Z. R., Wanyama, J. M., Midega, C. A., Pittchar, J., Hassanali, A., and Pickett, J. A. (2009). Evaluation of farmers' field days as a dissemination tool for push-pull technology in Western Kenya. *Crop Protection*, *28*(3), 225–235.

Anderson, J. R., Feder, G., and Ganguly, S. (2006). The rise and fall of training and visit extension: An Asian mini-drama with an African epilogue. World Bank Policy Research Working Paper. Washington, DC: World Bank.

Ashby, J., Kristjanson, P., Thornton, P., Campbell, B., Vermeulen, S., and Wollenberg, E. (2012). *CCAFS Gender Strategy*. Copenhagen, Denmark: CGIAR Research Program on Climate Change, Agriculture and Food Security (CCAFS).

Bahadur, A. V., Ibrahim, M., and Tanner, T. (2013). Characterising resilience: Unpacking the concept for tackling climate change and development. *Climate and Development*, *5*(1), 55–65.

Bernier, Q., and Meinzen-Dick, R. (2014). Social capital and resilience. In S. Fan, R. Pandya-Lorch and S. Yosef (Eds.), *Resilience for Food and Nutrition Ssecurity. An IFPRI 2020 book.* (pp. 169–178). Washington, DC: IFPRI.

Braun, A., Jiggins, J., Roling, N., van den Berg, H., and Snijders, P. (2006). *Global survey and review of farmer field school experiences.* Report prepared for the International Livestock Research Institute (ILRI). Wageningen: Endelea.

Brooks, S. and Loevinsohn, M. (2011). Shaping agricultural innovation systems responsive to food insecurity and climate change. *Natural Resources Forum*, *35*(3), 185–200.

Campbell, B. and Thornton, P. (2014). *How Many Farmers in 2030 and How Many Will Adopt Climate Resilient Innovations?* CCAFS Info Note. Copenhagen, Denmark: CGIAR Research Program on Climate Change, Agriculture and Food Security (CCAFS).

Chambers, R., and Jiggins, J. (1987). Agricultural research for resource-poor farmers part I: Transfer-of-technology and farming systems research. *Agricultural Administration and Extension*, *27*(1), 35–52.

Chipeta, S. (2013). *Gender Equality in Rural Advisory Services.* GFRAS Brief #2. Lindau, Switzerland: GFRAS.

Chowdhury, A. H., Hambly Odame, H., and Leeuwis, C. (2014). Transforming the roles of a public extension agency to strengthen innovation: Lessons from the national agricultural extension project in Bangladesh. *The Journal of Agricultural Education and Extension, 20*(1), 7–25.

Cooper, P., Dimes, J., Rao, K., Shapiro, B., Shiferaw, B., and Twomlow, S. (2008). Coping better with current climatic variability in the rain-fed farming systems of sub-Saharan Africa: An essential first step in adapting to future climate change? *Agriculture, Ecosystems and Environment, 126*(1), 24–35.

Cornwall, A., Guijt, I., and Welbourn, A. (1994). Extending the horizons of agricultural research and extension: Methodological challenges. *Agriculture and Human Values, 11*(2–3), 38–57.

Daniel, J. H., Lewis, L. W., Redwood, Y. A., Kieszak, S., Breiman, R. F., Flanders, W. D., Likimani, S. et al. (2011). Comprehensive assessment of maize aflatoxin levels in Eastern Kenya. *Environmental Health Perspectives, 119*(12), 1794–1799.

Darr, D. and Pretzsch, J. (2008). Mechanisms of innovation diffusion under information abundance and information scarcity—On the contribution of social networks in group vs. individual extension approaches in semi-arid Kenya. *Journal of Agricultural Education and Extension, 14*(3), 231–248.

Davis, K. (2008). Extension in sub-Saharan Africa: Overview and assessment of past and current models and future prospects. *Journal of International Agricultural and Extension Education, 15*(3), 15–28.

Davis, K., Babu, S. C., and Blom, S. (2014). Building the resilience of smallholders through extension and advisory services. In S. Fan, R. Pandya-Lorch and S. Yosef (Eds.), *Resilience for Food and Nutrition Security. An IFPRI 2020 book.* (pp. 127–135). Washington, DC: IFPRI.

Davis, K. and Heemskerk, W. (2012). Module 3: Investment in extension and advisory services as part of agricultural innovation systems. *Agricultural Innovation Systems: An Investment Sourcebook* (pp. 179–193). Washington, DC: World Bank.

Davis, K., Nkonya, E., Kato, E., Mekonnen, D. A., Odendo, M., Miiro, R., and Nkuba, J. (2012). Impact of farmer field schools on agricultural productivity and poverty in East Africa. *World Development, 40*(2), 402–413.

De Souza, K., Kituyi, E., Harvey, B., Leone, M., Murali, K. S., and Ford, J. D. (2015). Vulnerability to climate change in three hot spots in Africa and Asia: Key issues for policy-relevant adaptation and resilience-building research. *Regional Environmental Change, 15*(5), 747–753.

Deere, C. D. and Doss, C. R. (2006). The gender asset gap: What do we know and why does it matter? *Feminist Economics, 12*(1–2), 1–50.

Defoer, T., Wopereis, M. C. S., Idinoba, P., Kadisha, K. L., Diack, S., and Gaye, M. (2009). *Curriculum for Participatory Learning and Action Research (PLAR) for Integrated Rice Management (IRM) in Inland Valleys of Sub-Saharan Africa: Facilitator's Manual.* Cotonou, Benin: Africa Rice Center (WARDA).

DFID. (2007). *Gender Equality at the Heart of Development: Why the Role of Women Is Crucial to Ending World Poverty.* London, UK: DFID.

Doss, C. R. (2001). Designing agricultural technology for African women farmers: Lessons from 25 years of experience. *World Development, 29*(12), 2075–2092.

Dufour, C., Kauffmann, D., and Marsland, N. (2014). Enhancing the links between resilience and nutrition. In S. Fan, R. Pandya-Lorch and S. Yosef (Eds.), *Resilience for food and nutrition security. An IFPRI 2020 book.* (pp. 107–118). Washington, DC: IFPRI.

Eksvärd, K. and Björklund, J. (2010). Is PLAR (participatory learning and action research) a sufficient approach for the purpose of supporting transitions for sustainable agriculture?

A case study from Sweden. *Journal of Agricultural Extension and Rural Development, 2*(9), 179–190.

Ericksen, P. J. (2008). Conceptualizing food systems for global environmental change research. *Global Environmental Change, 18*(1), 234–245.

Evenson, R. E. and Mwabu, G. (2001). The effect of agricultural extension on farm yields in Kenya. *African Development Review, 13*(1), 1–23.

Fan, S., Pandya-Lorch, R., and Fritschel, H. (2012). Overview. In S. Fan and R. Pandya-Lorch (Eds.), *Reshaping Agriculture for Nutrition and Health. An IFPRI 2020 book.* (pp. 1–11). Washington, DC: IFPRI.

Fan, S., Pandya-Lorch, R., and Yosef, S. (2014). *Resilience for Food and Nutrition Security: An IFPRI 2020 book.* Washington, DC: IFPRI.

Farnworth, C. and Colverson, K. (2015). Building a gender-transformative extension and advisory facilitation system in Africa. *Journal of Gender, Agriculture and Food Security, 1*(1), 20–39.

Farrington, J. and Martin, A. M. (1988). Farmer participatory research: A review of concepts and recent fieldwork. *Agricultural Administration and Extension, 29*(4), 247–264.

Ferroni, M. and Castle, P. (2011). Public-private partnerships and sustainable agricultural development. *Sustainability, 3*, 1064–1073.

Fischer, E. and Qaim, M. (2012). Linking smallholders to markets: Determinants and impacts of farmer collective action in Kenya. *World Development, 40*(6), 1255–1268.

Foran, T., Butler, J. R., Williams, L. J., Wanjura, W. J., Hall, A., Carter, L., and Carberry, P. S. (2014). Taking complexity in food systems seriously: An interdisciplinary analysis. *World Development, 61*, 85–101.

Franzel, S., Sinja, J., and Simpson, B. (2014). *Farmer-to-Farmer Extension in Kenya: The Perspectives of Organizations using the Approach.* World Agroforestry Center Working Paper (Vol. 181). Nairobi, Kenya: Word Agroforestry Center.

Fraser, E. D., Dougill, A. J., Hubacek, K., Quinn, C. H., Sendzimir, J., and Termansen, M. (2011). Assessing vulnerability to climate change in dryland livelihood systems: Conceptual challenges and interdisciplinary solutions. *Ecology and Society, 16*(3), 3.

Friis-Hansen, E. and Duveskog, D. (2012). The empowerment route to well-being: An analysis of farmer field schools in East Africa. *World Development, 40*(2), 414–427.

Gichangi, E., Njiru, E., Itabari, J., Wambua, J., Maina, J., and Karuku, A. (2007). Assessment of improved soil fertility and water harvesting technologies through community based on-farm trials in the asals of Kenya. In A. Bationo (Ed.), *Advances in Integrated Soil Fertility Management in Sub-Saharan Africa: Challenges and Opportunities* (pp. 759–766). Dordrecht, The Netherlands: Springer.

Giller, K., Tittonell, P., Rufino, M. C., Van Wijk, M., Zingore, S., Mapfumo, P. et al. (2011). Communicating complexity: Integrated assessment of trade-offs concerning soil fertility management within African farming systems to support innovation and development. *Agricultural Systems, 104*(2), 191–203.

Graef, F., Sieber, S., Mutabazi, K., Asch, F., Biesalski, H. K., Bitegeko, J. et al. (2014). Framework for participatory food security research in rural food value chains. *Global Food Security, 3*(1), 8–15.

Griffith, A. and Osorio, L. (2008). *Participatory Market System Development: Best Practices in Implementation of Value Chain Development Programs.* MicroREPORT#149. USA: USAID and Practical Action.

Grisley, W. (1997). Crop-pest yield loss: A diagnostic study in the Kenya highlands. *International Journal of Pest Management, 43*(2), 137–142.

Guèye, E. (2000). The role of family poultry in poverty alleviation, food security and the promotion of gender equality in rural Africa. *Outlook on Agriculture, 29*(2), 129–136.

Haddad, L. (2013). From nutrition plus to nutrition driven: How to realize the elusive potential of agriculture for nutrition? *Food & Nutrition Bulletin, 34*(1), 39–44.

Hall, A., Janssen, W., and Rajalahti, R. (2006). *Enhancing agricultural innovation: How to Go beyond The Strengthening of Research Systems.* Washington, DC: World Bank.

Hawkes, C. and Ruel, M.T. (2012). Value chains for nutrition. In S. Fan and R. Pandya-Lorch (Eds.), *Reshaping agriculture for nutrition and health. An IFPRI 2020 book* (pp. 73–82). Washington, DC: IFPRI.

Herforth, A., Lidder, P., and Gill, M. (2015). Strengthening the links between nutrition and health outcomes and agricultural research. *Food Security, 7*(3), 457–461.

IFAD. (2012). *Gender Equality and Women's Empowerment: Policy.* Rome, Italy: IFAD.

Ifejika Speranza, C. (2006). Gender-based analysis of vulnerability to drought among agro-pastoral households in semi-arid Makueni district, Kenya. In S. Premchander and C. Müller (Eds.), *Gender and Sustainable Development. Case Studies from NCCR North–South* (Vol. 2, pp. 119–146). Bern: Geographica Bernensia.

Jaetzold, R., Schmidt, H., Hornetz, B., and Shisanya, C. (2006). *Farm Management Handbook of Kenya vol. I—Natural Conditions and Farm Management Information,* 2nd edition part C East Kenya subpart c1 Eastern Province. Nairobi, Kenya: Ministry of Agriculture.

Jafry, T. and Sulaiman, V. R. (2013). Gender-sensitive approaches to extension programme design. *Journal of Agricultural Education and Extension, 19*(5), 469–485.

Jayne, T., Chamberlin, J., and Headey, D. D. (2014). Land pressures, the evolution of farming systems, and development strategies in Africa: A synthesis. *Food Policy, 48,* 1–17.

Johns, T., Powell, B., Maundu, P., and Eyzaguirre, P. B. (2013). Agricultural biodiversity as a link between traditional food systems and contemporary development, social integrity and ecological health. *Journal of the Science of Food and Agriculture, 93*(14), 3433–3442.

Kabeer, N., and Natali, L. (2013). *Gender Equality and Economic Growth: Is There a Win–Win?* IDS Working Paper 417. Brighton, UK: International Development Studies.

Kilelu, C. W., Klerkx, L., Leeuwis, C., and Hall, A. (2011). Beyond knowledge brokering: An exploratory study on innovation intermediaries in an evolving smallholder agricultural system in Kenya. *Knowledge Management for Development Journal, 7*(1), 84–108.

Kingiri, A. N. (2013). A review of innovation systems framework as a tool for gendering agricultural innovations: Exploring gendered learning and systems empowerment. *Journal of Agricultural Education and Extension, 19*(5), 521–541.

Kiptot, E. and Franzel, S. (2012). Gender and agroforestry in Africa: Who benefits? The African perspective. In P. K. R. Nair and D. Garrity (Eds.), *Agroforestry—The Future of Global Land Use* (pp. 463–496). Dordrecht, The Netherlands: Springer.

Kiptot, E., and Franzel, S. (2015). Farmer-to-farmer extension: Opportunities for enhancing performance of volunteer farmer trainers in Kenya. *Development in Practice, 25*(4), 503–517.

Klerkx, L., Schut, M., Leeuwis, C., and Kilelu, C. (2012). Advances in knowledge brokering in the agricultural sector: Towards innovation system facilitation. *IDS Bulletin, 43*(5), 53–60.

Klerkx, L., Van Mierlo, B., and Leeuwis, C. (2012). Evolution of systems approaches to agricultural innovation: Concepts, analysis and interventions. In I. Darnhofer, D. Gibbon and B. Dedieu (Eds.), *Farming Systems Research into the 21st Century: The New Dynamic* (pp. 457–483). Dordrecht, The Netherlands: Springer.

Kristjanson, P., Harvey, B., Van Epp, M., and Thornton, P. K. (2014). Social learning and sustainable development. *Nature Climate Change, 4*(1), 5–7.

Kumar, N. and Quisumbing, A. (2014). Gender and resilience. In S. Fan, R. Pandya-Lorch and S. Yosef (Eds.), *Resilience for Food and Nutrition Security: An IFPRI 2020 book.* (pp. 155–167). Washington, DC: IFPRI.

Larsen, A. F. and Lilleør, H. B. (2014). Beyond the field: The impact of farmer field schools on food security and poverty alleviation. *World Development, 64,* 843–859.

Leeuwis, C. and Aarts, N. (2011). Rethinking communication in innovation processes: Creating space for change in complex systems. *The Journal of Agricultural Education and Extension, 17*(1), 21–36.

Leeuwis, C., Hall, A., van Weperen, W., and Preissing, J. (2013). *Facing the Challenges of Climate Change and Food Security: The role of Research, Extension and Communication for Development*. Rome, Italy: FAO.

Lim, S. S., Winter-Nelson, A., and Arends-Kuenning, M. (2007). Household bargaining power and agricultural supply response: Evidence from Ethiopian coffee growers. *World Development*, *35*(7), 1204–1220.

Lipper, L., Thornton, P., Campbell, B. M., Baedeker, T., Braimoh, A., Bwalya, M. et al. (2014). Climate-smart agriculture for food security. *Nature Climate Change*, *4*(12), 1068–1072.

Magothe, T., Okeno, T. O., Muhuyi, W., and Kahi, A. (2012). Indigenous chicken production in Kenya: I. Current status. *World's Poultry Science Journal*, *68*(01), 119–132.

Mailu, S. K., Wachira, M. A., Munyasi, J. W., Nzioka, M., Kibiru, S. K., Mwangi, D. M. et al. (2012). Influence of prices on market participation decisions of indigenous poultry farmers in four districts of Eastern Province, Kenya. *Journal of Agriculture and Social Research*, *12*(1), 1–10.

Maina, I., Newsham, A., and Okoti, M. (2013). *Agriculture and Climate Change in Kenya: Climate Chaos, Policy Dilemmas*. Working Paper 070. Brighton, UK: Future Agricultures Consortium.

Markelova, H., Meinzen-Dick, R., Hellin, J., and Dohrn, S. (2009). Collective action for smallholder market access. *Food Policy*, *34*(1), 1–7.

Mayoux, L. (2000). *Micro-Finance and the Empowerment of Women: A Review of the Key Issues*. Geneva, Switzerland: International Labour Office, Employment Sector, Social Finance Unit.

Mbo'o-Tchouawou, M. and Colverson, K. (2014). *Increasing Access to Agricultural Extension and Advisory Services: How Effective Are New Approaches in Reaching Women Farmers in Rural Areas?* Nairobi, Kenya: ILRI.

Mbow, C., Van Noordwijk, M., Luedeling, E., Neufeldt, H., Minang, P. A., and Kowero, G. (2014). Agroforestry solutions to address food security and climate change challenges in Africa. *Current Opinion in Environmental Sustainability*, *6*, 61–67.

Meinzen-Dick, R., Behrman, J., Menon, P., and Quisumbing, A. (2012). Gender: A key dimension linking agricultural programs to improved nutrition and health. In S. Fan and R. Pandya-Lorch (Eds.), *Reshaping Agriculture for Nutrition and Health. An IFPRI 2020 book.* (pp. 135–144). Washington, DC: IFPRI.

Meinzen-Dick, R., Johnson, N., Quisumbing, A., Njuki, J., Behrman, J., Rubin, D., Waithanji, E. et al. (2011). *Gender, Assets, and Agricultural Development Programs: A Conceptual Framework*. CAPRi Working Paper No. 99. Washington, DC: IFPRI.

Meinzen-Dick, R., Quisumbing, A., Behrman, J., Biermayr-Jenzano, P., Wilde, V., Noordeloos, M., Beintema, N. et al. (2011). *Engendering Agricultural Research, Development and Extension*. Washington, DC: IFPRI.

Mganga, K. Z., Musimba, N. K. R., Nyariki, D. M., Nyangito, M. M., and Mwang'ombe, A. W. (2015). The choice of grass species to combat desertification in semi-arid Kenyan rangelands is greatly influenced by their forage value for livestock. *Grass and Forage Science*, *70*(1), 161–167.

Miruka, M. K., Okello, J. J., Kirigua, V. O., and Murithi, F. M. (2012). The role of the Kenya Agricultural Research Institute (KARI) in the attainment of household food security in Kenya: A policy and organizational review. *Food Security*, *4*(3), 341–354.

Misselhorn, A., Aggarwal, P., Ericksen, P., Gregory, P., Horn-Phathanothai, L., Ingram, J., and Wiebe, K. (2012). A vision for attaining food security. *Current Opinion in Environmental Sustainability*, *4*(1), 7–17.

Muhammad, L., Njoroge, K., Bett, C., Mwangi, W., Verkuijl, H., and De Groote, H. (2003). *The Seed Industry for Dryland Crops in Eastern Kenya*. Mexico: CIMMYT.

Mungube, E. O., Bauni, S. M., Tenhagen, B. A., Wamae, L. W., Nzioka, S. M., Muhammed, L., and Nginyi, J. M. (2008). Prevalence of parasites of the local scavenging chickens in a selected semi-arid zone of Eastern Kenya. *Tropical Animal Health and Production*, *40*(2), 101–109.

Murray-Prior, R. (2013). Developing an agricultural innovation system to meet the needs of smallholder farmers in developing countries. *Extension Farming Systems Journal, 9*(1), 258–263.

Muthoni, J., and Nyamongo, D. (2008). Seed systems in Kenya and their relationship to on-farm conservation of food crops. *Journal of New Seeds, 9*(4), 330–342.

Muyanga, M. and Jayne, T. (2008). Private agricultural extension system in Kenya: Practice and policy lessons. *Journal of Agricultural Education and Extension, 14*(2), 111–124.

Nagarajan, L., Audi, P., Jones, R., and Smale, M. (2007). *Seed Provision and Dryland Crops in the Semiarid Regions of Eastern Kenya.* Washington, DC: IFPRI.

Najjar, D., Spaling, H., and Sinclair, A. J. (2013). Learning about sustainability and gender through farmer field schools in the Taita Hills, Kenya. *International Journal of Educational Development, 33*(5), 466–475.

Nelson, V. and Stathers, T. (2009). Resilience, power, culture, and climate: A case study from semi-arid Tanzania, and new research directions. *Gender & Development, 17*(1), 81–94.

Njuguna, E., Brownhill, L., Kihoro, E., Muhammad, L. W., and Hickey, G. M. (in press). Gendered technology adoption and household food security in semi-arid Eastern Kenya. In J. Parkins, J. Njuki and A. Kaler (Eds.), *Transforming Gender and Food Security in the Global South.* London, UK: Earthscan.

Njuki, J., Kaaria, S., Chamunorwa, A., and Chiuri, W. (2011). Linking smallholder farmers to markets, gender and intra-household dynamics: Does the choice of commodity matter? *European Journal of Development Research, 23*(3), 426–443.

Nyariki, D. M., Wiggins, L. S., and Imungi, J. K. (2002). Levels and causes of household food and nutrition insecurity in dryland Kenya. *Ecology of Food and Nutrition, 41*(2), 155–176.

Ogada, M., Muchai, D., Mwabu, G., and Mathenge, M. (2014). Technical efficiency of Kenya's smallholder food crop farmers: Do environmental factors matter? *Environment, Development and Sustainability, 16*(5), 1065–1076.

Okeno, T. O., Kahi, A. K., and Peters, K. J. (2012). Characterization of indigenous chicken production systems in Kenya. *Tropical Animal Health and Production, 44*(3), 601–608.

Pant, L. P. and Hambly-Odame, H. (2009). Innovations systems in renewable natural resource management and sustainable agriculture: A literature review. *African Journal of Science, Technology, Innovation and Development, 1*(1), 103–135.

Poulton, C. and Kanyinga, K. (2014). The politics of revitalising agriculture in Kenya. *Development Policy Review, 32*(s2), s151–s172.

Poulton, C. and Macartney, J. (2012). Can public–private partnerships leverage private investment in agricultural value chains in Africa? A preliminary review. *World Development, 40*(1), 96–109.

Ragasa, C., Berhane, G., Tadesse, F., and Taffesse, A. S. (2013). Gender differences in access to extension services and agricultural productivity. *Journal of Agricultural Education and Extension, 19*(5), 437–468.

Reijntjes, C., Haverkort, B., and Waters Bayer, A. (1992). *Farming for the Future: An Introduction to Low-External-Input and Sustainable Agriculture.* London, UK: Macmillan.

Republic of Kenya. (2013). *National Climate Change Action Plan 2013–2017.* Nairobi: Republic of Kenya.

Rivera, W. M. and Qamar, M. K. (2003). *Agricultural Extension, Rural Development and the Food Security Challenge.* Rome, Italy: FAO.

Rivera, W. M. and Sulaiman, V. R. (2009). Extension: Object of reform, engine for innovation. *Outlook on Agriculture, 38*(3), 267–273.

Rocheleau, D. (1999). Confronting complexity, dealing with difference: Social context, content and practice in agroforestry. In L. E. Buck, J. P. Lassoie and E. C. M. Fernandes (Eds.), *Agroforestry in Sustainable Agricultural Systems* (pp. 191–235). Boca Raton: CRC Press LLC.

Rogers, E. M. (1962). *Diffusion of Innovations.* New York, NY: Free Press of Glencoe.

Röling, N. (2009). Conceptual and methodological developments in innovation. In P. Sanginga, A. Waters-Bayer, S. Kaaria, J. Njuki and C. Wettasinha (Eds.), *Innovation Africa: Enriching Farmers' Livelihoods* (pp. 9–34). London, UK: Earthscan.

Sanginga, P., Waters-Bayer, A., Kaaria, S., Njuki, J., and Wettasinha, C. (2009). Innovation Africa: Beyond rhetoric to praxis. In P. Sanginga, A. Waters-Bayer, S. Kaaria, J. Njuki and C. Wettasinha (Eds.), *Innovation Africa: Enriching Farmers' Livelihoods* (pp. 374–386). London, UK: Earthscan.

Shiferaw, B., Hellin, J., and Muricho, G. (2011). Improving market access and agricultural productivity growth in Africa: What role for producer organizations and collective action institutions? *Food Security, 3*(4), 475–489.

Shiferaw, B., Tesfaye, K., Kassie, M., Abate, T., Prasanna, B. M., and Menkir, A. (2014). Managing vulnerability to drought and enhancing livelihood resilience in sub-Saharan Africa: Technological, institutional and policy options. *Weather and Climate Extremes, 3*, 67–79.

Shumsky, S. A., Hickey, G. M., Pelletier, B., and Johns, T. (2014). Understanding the contribution of wild edible plants to rural social-ecological resilience in semi-arid Kenya. *Ecology and Society, 19*(4), 34.

Sietz, D., Lüdeke, M. K., and Walther, C. (2011). Categorisation of typical vulnerability patterns in global drylands. *Global Environmental Change, 21*(2), 431–440.

Smith, J. (2015). *Crops, Crop Pests and Climate Change—Why Africa Needs to be Better Prepared.* Working Paper No. 114. Copenhagen, Denmark: CGIAR Research Program on Climate Change, Agriculture and Food Security (CCAFS).

Snapp, S. and Heong, K. (2003). Scaling up and out. In B. Pound, S. Snapp, C. McDougall and A. Braun (Eds.), *Managing Natural Resources for Sustainable Livelihoods: Uniting Science and Participation* (pp. 67–87). London, UK: Earthscan.

Stathers, T., Lamboll, R., and Mvumi, B. M. (2013). Postharvest agriculture in changing climates: Its importance to African smallholder farmers. *Food Security, 5*(3), 361–392.

Struik, P. C., Klerkx, L., van Huis, A., and Röling, N. G. (2014). Institutional change towards sustainable agriculture in West Africa. *International Journal of Agricultural Sustainability, 12*(3), 203–213.

Sulaiman, R. and Davis, K. (2012). *The "New Extensionist": Roles, Strategies, and Capacities to Strengthen Extension and Advisory Services.* Lindau, Switzerland: Global Forum for Rural Advisory Services.

Sustainet EA. (2010). *Technical Manual for Farmers and Field Extension Service Providers: Farmer Field Schoolaapproach.* Nairobi, Kenya: Sustainable Agriculture Information Initiative.

Thornton, P., Van de Steeg, J., Notenbaert, A., and Herrero, M. (2009). The impacts of climate change on livestock and livestock systems in developing countries: A review of what we know and what we need to know. *Agricultural Systems, 101*(3), 113–127.

Tittonell, P. (2014). Livelihood strategies, resilience and transformability in African agroecosystems. *Agricultural Systems, 126*, 3–14.

Uphoff, N. (2014). Systems thinking on intensification and sustainability: Systems boundaries, processes and dimensions. *Current Opinion in Environmental Sustainability, 8*, 89–100.

van den Berg, H. and Jiggins, J. (2007). Investing in farmers—the impacts of farmer field schools in relation to integrated pest management. *World Development, 35*(4), 663–686.

van Ginkel, M., Sayer, J., Sinclair, F., Aw-Hassan, A., Bossio, D., Craufurd, P. et al. (2013). An integrated agro-ecosystem and livelihood systems approach for the poor and vulnerable in dry areas. *Food Security, 5*(6), 751–767.

WARDA. (2003). *Annual Report 2002–2003.* Cotonou, Benin: WARDA (Africa Rice Centre).

World Bank. (2007). *World Development Report 2008: Agriculture for Development.* Washington, DC: World Bank.

World Bank and IFPRI. (2010). *Gender and Governance in Rural Services: Insights from India, Ghana and Ethiopia.* Gender and Governance Author Team. Washington, DC: World Bank.

Zorya, S., Morgan, N., Diaz Rios, L., Hodges, R., Bennett, B., Stathers, T. et al. (2011). *Missing food: The Case of Postharvest Grain Losses in Sub-Saharan Africa.* Technical Report. Washington, DC: World Bank.

3 Exploring the relationships between gender, social networks, and agricultural innovation in two smallholder farming communities in Machakos County, Kenya

Colleen M. Eidt, Gordon M. Hickey, and Bernard Pelletier

Introduction

Within the context of integrated human and natural (or social-ecological) systems, resilience has been defined as the ability of a system to regain full function following a disturbance (Holling, 1973). This definition incorporates both the possibility of returning to an existing state (or states) and the potential for adaptation, social learning, and change (Adger, Hughes, Folke, Carpenter, and Rockström, 2005; Cutter et al., 2008; Folke, 2006). A wide range of factors contribute to the resilience of a social-ecological system, resulting in divergent frameworks for measuring related indicators and a lack of attention to the social dynamics of socio-ecological systems (Bahadur, Ibrahim, and Tanner, 2013), including issues of power, politics, and agency (Bahadur and Tanner, 2014; Cote and Nightingale, 2012; Turner, 2010). In particular, efforts to foster social-ecological resilience need to explicitly address the importance of authority and power in order to avoid the dangers of elite capture and the further exclusion and marginalization of already marginalized groups (Newman and Dale, 2005; Swanstrom, 2008), particularly women.

Women living in rural farming communities throughout Sub-Saharan Africa are known to have less access to agricultural resources and employment than their male counterparts, which has negative implications for human development, agricultural output, and household food security (FAO, 2011). Inequalities in divisions of labour, ability to participate in decision making, and access to resources also shape men's and women's adaptive capacities (Rossi and Lambrou, 2008). Feminist scholars argue that in order to understand this intersection between gender and adaptation to change, the discourse must move beyond measuring differences in outcomes (Tuana and Cuomo, 2014) and acknowledge that the silencing of women's voices effectively prioritises other groups and renders women more vulnerable (Banford and Froude, 2015). Yet rural women, as central carriers of indigenous knowledge and traditional farming practices, are key to building resilience (Brownhill, 2010).

Gender-sensitive research is now recognized as essential within the social-ecological resilience framework in order to better understand how potential shifts within a system may affect individuals differently based on gender (Nelson and Stathers, 2009) and how gender affects capabilities to foster resilience. If theories of resilience are blind to gender, then the research, policy, and programs informed by these theories may undermine efforts to promote gender empowerment and equity (Nelson and Stathers, 2009) as well as potentially undermine the resilience initiatives themselves. This chapter explores the gendered nature of the social networks operating in two communities in semi-arid Eastern Kenya and reflects on some of the implications for agricultural system innovation in support of resilience. Importantly, there has been a long history of participatory agricultural interventions (Cornwall, 2008) within the semi-arid regions of Kenya (Tiffen, Mortimore, and Gichuki, 1994). Participation is thought to empower local communities, through networking and capacity building, and also to make interventions more efficient in delivering outcomes (Pretty, 1995). There is, however, an inconsistent track record in the perceived success or failure of these types of interventions (Reed, 2008), often resulting in significantly different legacies and levels of innovation potential in communities.

Research design

Study setting

This research focused on small-scale farming systems in Kenya's semi-arid eastern counties, where hunger and malnutrition are prevalent. The area is characterised by two growing periods coinciding with the long (March-April-May) and short (October-November-December) rainy seasons. The soils range from clay-based to sandy and are often rocky, particularly in areas around the Yatta Plateau, Machakos County, where our study sites are located. Staple crops in the region include cereals (maize, sorghum, and millet), pulses (beans, pigeon peas, cowpeas, and green grams), and roots and tubers (sweet potato, cassava, yam, and arrowroot) (Hickey, Pelletier, Brownhill, Kamau, and Maina, 2012; Government of Kenya, 2010). Small-scale farming, rain-fed agriculture, and low levels of technology adoption characterise these semi-arid regions (Miruka, Okello, Kirigua, and Murithi, 2012). Communities are vulnerable to shocks from droughts and frequent crop failure, food insecurity, and heavy reliance on food aid (KFSSG, 2013, 2014).

Case selection, data collection, and analysis

Through consultations with project partners at the Kenyan Agricultural and Livestock Research Organisation (KALRO), a study location was identified in Yatta sub-county, Machakos County. Yatta has a long history of agricultural interventions that have been funded and implemented by many different stakeholder groups, using a variety of approaches and experiencing different levels of success. These range from large-scale, internationally funded programs carried out by government employees to the efforts of both local and international individuals to foster grassroots change. For example, the National Agricultural and Livestock Extension Program (NALEP), implemented by the Ministry of Agriculture and funded by the Government of Kenya and the Swedish International Development Agency, were operating in Yatta using a 'demand-driven' approach. Under this program farmers organized into groups to select and implement a range of activities, including options such as rearing dairy goats or planting local vegetables to improve their agricultural productivity. On a smaller scale, local leaders in Yatta have also formed grassroots NGOs that deliver training and support to farmers concerning a range of farming challenges such as irrigation and water harvesting technologies.

To select the two villages to be included in the study, a pair-wise ranking of the sub-locations in Yatta was carried out by a panel of local experts including NGO workers, researchers, and agricultural extension officers, using the following criteria: 1) located within the semi-arid Agro-Ecological Zone (AEZ) Lower Midlands (LM) 4 or 5, as opposed to areas of higher agricultural potential located in AEZ Upper Midlands (UM) 4, depicted in Figure 3.1; 2) a history of multiple agricultural interventions ranging in levels of perceived success; and 3) presence of a range of active stakeholders (agricultural extension officers, researchers, farmers, etc.) representing different groups (non-government organisations, or NGOs, government organisation, private industry, etc.). This ranking exercise was dependent on the panel's understanding of what defines 'intervention success' in terms of building resilience, namely meeting certain development targets they identified, such as increasing on-farm biodiversity, the application of new technologies, and moving beyond relying solely on rain-fed agriculture. In this study, we used the perspective of an 'on the ground' project-implementing stakeholders as a proxy for determining innovation potential. The panel collectively identified Mathingau as the most developed and innovative sub-location and Kyua as the least. Finally, in collaboration with local leaders (chiefs, assistant chiefs, agricultural extension officers, etc.), one village within each sub-location was selected that best exemplified the relatively high and low levels of agricultural development and smallholder innovation found in Mathingau and Kyua, respectively.

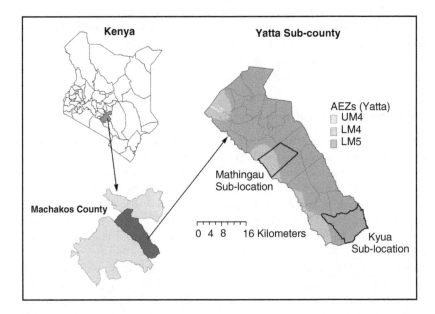

Figure 3.1 Map of the study area.

Table 3.1 Household survey and dissemination workshop participants.

Study	Participants (N)	
	Mathingau	Kyua
Household survey	72 (47 women, 25 men)	73 (46 women, 27 men)
Dissemination workshops	Approximately 100	Approximately 75

A complete household survey was then conducted in July 2013, sampling every male and female head of household within each selected village (see Table 3.1). The survey was first pre-tested with a group of farmers and then revised. Questions were multiple-choice and focused on factors associated with social resilience (i.e., education, coping strategies, knowledge networks, participation, and empowerment) (see Bahadur et al., 2013). Participation in this context was measured only through involvement in any type of farmer's organisation, co-op, or group. We acknowledge the limitation to this approach, as groups may be inactive or membership may not indicate 'genuine' participation. Nevertheless, we suggest that group membership is a relevant indicator of some form of participation or at least willingness to participate and therefore it served the purpose of our study. Also, self-reported ability to make potentially life altering decisions was used as a proxy for empowerment. Survey data were analysed using Chi square (X^2) tests of independence conducted in Stata to explore differences between villages and genders. Finally, in July 2014, a follow-up dissemination workshop was conducted in each sub-location to discuss the emerging results of the study to the participating communities, which helped to ensure face validity (Rubin and Babbie, 2007).

Results

Food insecurity and education

There was a significant association between self-reported household food insecurity and village (p=0.001), with food insecurity being more prevalent in Kyua (86.3%), where past agricultural interventions are thought to have been less successful than in Mathingau (62.5%). However, when broken down by gender, this association weakens for men (p=0.156) and continues to be significant for women (p=0.002). This suggests that previous successful interventions in Mathingau may have had more impact on improving women's food security when compared with the levels found among women in Kyua, while levels of food security among men remained consistent between the two villages. There were also significant links between lower levels of education and food insecurity (p<0.001) as well as education and village (p=0.030), with individuals in Kyua, regardless of gender, possessing lower levels of formal education. There were no gender differences in education measures, other than an association between achieving at least a secondary education and being from Mathingau among men (p=0.033).

When determining how individuals cope in times of food insecurity, there was a significant relationship between relying on friends, family, or neighbours and village, with this being a more common coping strategy in Kyua (41.3%) than in Mathingau (20.0%) (p=0.020). When broken down by gender, irrespective of village, this trend disappears for men (p=0.647) and remains significant for women (p=0.012). Other coping mechanisms such as borrowing from a formal lending institution did not differ between communities or genders.

Knowledge networks

When asked to select the most important source of agricultural knowledge, significant differences were observed between villages (p<0.001) that were consistent for men and women (p<0.001 in both cases). Both men and women from Mathingau selected a wider range of sources, with NGOs (29.2%); family, friends, or neighbours (23.6%); or government agricultural extension (19.4%) being the most frequent responses. In Kyua participants relied mostly on the media (48.6%) and overall displayed less diversity in responses. Of particular importance is the fact that government agricultural extension and NGOs, two key extension providers in rural areas of Kenya were selected by only 5.5% and 0% of Kyua respondents, respectively. Furthermore, within Kyua there were significant differences between the responses of men and women (p=0.026). While men selected media and family, friends, and neighbours equally (40.7%), women as a group were more diverse in their answers, selecting media most frequently (53.3%), followed by other sources and then family, friends, and neighbours much less frequently than men. This suggests that a history of successful interventions in Mathingau may have contributed to diversifying knowledge sources, particularly those of men, as in Kyua men, as a group, were less diverse in their responses than women. However, this may have also resulted in men not selecting family, friends, and neighbours as the most important source of agricultural knowledge as often as in Mathingau.

Participation and empowerment

There was a significant association between membership in some type of farmer's organisation, co-op, or group and village, with those in Mathingau more commonly holding some form of membership than those in Kyua (88.79% vs. 37.0%, p<0.001). This relationship remained consistent among men and women (p<0.001 in both cases). There is also a link between gender and participation, with women more likely to belong to a group than men (71.7% vs. 46.1%, p=0.002) regardless of village. When broken down by village, this trend remains in Kyua (p=0.003) and to a lesser extent in Mathingau (p=0.086). With regards to self-reported empowerment, participants were asked about their ability to make decisions which would change or improve their lives. The levels of empowerment were relatively high among men (82.69%) and women (81.72%). There were no significant relationships between villages or genders.

The comparison of resilience factors between the two villages, and the associations between these resilience factors and village by gender, are summarised in Tables 3.2 and 3.3, respectively.

Feedback from dissemination workshops

At the dissemination workshops, participating smallholder and subsistence farmers from both communities generally agreed with the survey findings. They also raised important issues and gave insights into some of the reasons why innovation potential varied between the communities. For example, there was a very active and successful community-based NGO working in Mathingau as well as provincial administrators who were generally perceived as being involved in and committed to improving the lives of the people within their communities. Farmers from both communities felt that the trust and legitimacy of these organisations and individuals contributed significantly to the success of previous agricultural interventions within Mathingau. Individuals in Kyua also pointed to past agricultural

Table 3.2 Comparison of resilience factors between representative villages in Mathingau (more innovative) and Kyua (less innovative)

Resilience Factor	Village	
	Mathingau	*Kyua*
Food Security	+	−
Education	+	−
Coping Strategies		
Reliance on family, friends, or neighbours	−	+
Other	=	=
Knowledge Networks		
Reliance on family, friends, or neighbours	−	+
Media	−	+
NGOs	+	−
Government agricultural extension	+	−
Participation	+	−
Empowerment	=	=

(+, −, and = represent relative differences)

Table 3.3 Comparison of resilience factor independence or association with village by gender

Resilience Factor	Gender	
	Men	Women
Food Security	Independent of village	Associated with village
Education	Independent of village	Independent of village
Coping Strategies		
Reliance on family, friends, or neighbours	Independent of village	Associated with village
Knowledge Networks	Associated with village	Associated with village
Participation	Associated with village	Associated with village
Empowerment	Independent of village	Independent of village

interventions involving the introduction of new technologies that failed to meet their expectations, creating distrust and a lack of interest in future projects. However, individuals within Mathingau reported similar let-downs but remain dedicated to participating in agricultural innovation programs, most likely due, at least in part, to the encouragement and support of trusted community leaders (including extension officers, NGO workers, and researchers) who were able to assist in the alignment of community and project expectations and objectives and as a result build trust and legitimacy.

Discussion

Collective action and social learning are central to a community's ability to adapt to environmental change and therefore are a key component of resilience (Gunderson, 2001). Strong networks and relations of trust facilitate collective problem solving (Mayunga, 2007) and social learning by giving community members access to coping strategies that are beyond their range of everyday experience (Tompkins and Adger, 2004). Subsequently, when networks and trust are missing, the community tends to have less adaptive capacity (Mayunga, 2007; Lowitt, Hickey, Ganpat, and Phillip, 2015). In the context of the two communities we studied in semi-arid Kenya, we can see that social networks are serving two distinct functions. Firstly, they serve as an important coping strategy in times of food shortages when individuals rely on family, friends, and neighbours for assistance. Secondly, they act as communication networks, transferring agricultural knowledge between individuals and organisations.

During times of food insecurity, men often have a wider range of coping strategies available to them than women. For example, they have greater mobility, an easier time securing off-farm employment, and often maintain the power to buy and sell livestock (FAO, 2011). Women, in contrast, often depend more heavily on social relationships in times of hardship and as a result are thought to better maintain their networks, which are distinct from men's (Agarwal, 2000). This may partially explain why we found participation through some form of group membership higher among women than men. However, women in Mathingau did not rely as heavily on their social networks as a coping strategy in times of food insecurity, suggesting that they likely had more formal options for accessing resources through the local agricultural innovation system. In contrast, women in Kyua continued to rely heavily on their social networks, suggesting a more informal and localized (and potentially closed) innovation system. These differences may go some way towards accounting for differences in household food security observed between the communities, and particularly women in each community.

Power is often tied to control over different forms of capital, including human, physical, natural, financial, and social, which men and women hold in different amounts based, to some extent, on traditional divisions of labour (Agarwal, 2000). Meeting the nutritional needs of the household in semi-arid Kenya is generally the woman's role (FAO, 2011) and by relying on their social networks to meet this need, women are using their social capital. This use of social capital is a potential source of power within the household which can often be affected by a shift to more formal and market-based arrangements requiring, for example, financial capital. Bahadur and Tanner (2014) have highlighted the importance of considering the implications of these types of trade-offs in operationalizing resiliency programs in order to prevent further marginalisation, in this case, of women.

The results of our study support this view, indicating that women farmers embedded in a more successful or effective agricultural innovation system (Mathingau) were less dependent on families, friends, and neighbours for knowledge and support during times of stress. Our results also revealed that men in Mathingau had more diversified sources of agricultural knowledge and less reliance on their family, friends, and relatives as sources of agricultural knowledge than did men in Kyua. This difference between the communities can be thought of as a shift from a reliance on social networks consisting of 'bonding ties' (connections that link similar individuals, such as family, friends, or neighbours) to one based more on 'bridging ties' (linking diverse individuals, such as those from different stakeholder groups) (Putnam, 2001). This shift is important in the context of agricultural innovation systems as networks that consist of only bonding ties tend to foster group homophily, reducing the potential for innovation and resilience (Newman and Dale, 2005), while bridging ties can give community members access to new technical and experiential knowledge, which would otherwise be beyond their reach (Tompkins and Adger, 2004).

Additionally, the resources (human, physical, natural, financial, and social) embedded in different network structures can be significant for adaptive capacity and therefore to resilience (Gidengil and Stolle, 2009). Norris, Stevens, Pfefferbaum, Wyche, and Pfefferbaum (2008) emphasise the importance of resources, suggesting that resilience is a factor of linking adaptive capacities through networks. It is likely that connecting more diverse individuals (for example, a farmer with an NGO worker) through bridging ties decreases homophily and introduces a greater diversity of resources into the network, including various forms of capital and adaptive capacity. Our results support this observation and highlight the need for agricultural initiatives to work towards strengthening the social capital of women and men smallholder farmers (Bernier and Meinzen-Dick, 2014).

Conclusion

There are many factors which contribute to the socio-ecological resilience of a system. Here we have considered only a few—education, coping strategies, knowledge networks, participation, and empowerment—in order to explore how they compare in different agricultural innovation system settings. Our results highlight the important and gendered roles that networks play in fostering innovation potential in support of social-ecological resilience. First, they highlight that social networks can be a valuable resource to women in times of food insecurity, allowing them to draw on their social capital and potentially contribute to their household power in relation to men. Agricultural interventions which alter women's coping strategies, while increasing access to other forms of capital, need to carefully consider these trade-offs. Second, we suggest that the potential of agricultural innovation systems will likely be positively influenced by shifting men's and women's knowledge networks away from being dependent on predominantly bonding ties and towards more bridging ties, a transformation likely to increase their access to various forms of capital and foster agricultural innovation. However, how this shift may impact household gender relationships requires further investigation. Our study offers some useful insights into highly complex relationships between men, women, social networks, and agricultural innovation in the rural smallholder communities of semi-arid Eastern Kenya.

References

Adger, W. N., Hughes, T. P., Folke, C., Carpenter, S. R., and Rockström, J. (2005). Social-ecological resilience to coastal disasters. *Science, 309*(5737), 1036–1039.

Agarwal, B. (2000). Conceptualising environmental collective action: Why gender matters. *Cambridge Journal of Economics, 24*(3), 283–310.

Bahadur, A. and Tanner, T. (2014). Transformational resilience thinking: Putting people, power and politics at the heart of urban climate resilience. *Environment and Urbanization, 26*(1), 200–214.

Bahadur, A. V., Ibrahim, M., and Tanner, T. (2013). Characterising resilience: Unpacking the concept for tackling climate change and development. *Climate and Development, 5*(1), 55–65.

Banford, A. and Froude, C. K. (2015). Ecofeminism and natural disasters: Sri Lankan women post-tsunami. *Journal of International Women's Studies, 16*(2), 170–187.

Bernier, Q. and Meinzen-Dick, R. (2014). Social capital and resilience. In S. Fan, R. Pandya-Lorch and S. Yosef (Eds.), *Resilience for Food and Nutrition Security. An IFPRI 2020 book.* (pp. 169–178). Washington, DC: IFPRI.

Brownhill, L. (2010) Gendered struggles for the commons: Food sovereignty, tree-planting and climate change. In J. Lee and S. Shaw (Eds), *Women Worldwide: Transnational Feminist Perspectives on Women* (pp. 484–487). Columbus, OH: McGraw-Hill Higher Education.

Cornwall, A. (2008). Unpacking 'Participation': Models, meanings and practices. *Community Development Journal*, *43*(3), 269–283.

Cote, M. and Nightingale, A. J. (2012). Resilience thinking meets social theory: Situating social change in socio-ecological systems (SES) research. *Progress in Human Geography*, *36*(4), 475–489.

Cutter, S. L., Barnes, L., Berry, M., Burton, C., Evans, E., Tate, E., and Webb, J. (2008). A place-based model for understanding community resilience to natural disasters. *Global Environmental Change*, *18*(4), 598–606.

FAO. (2011). *The State of Food and Agriculture 2010–2011. Women in Agriculture. Closing the Gender Gap for Development*. Rome, Italy: FAO.

Folke, C. (2006). Resilience: The emergence of a perspective for social–ecological systems analyses. *Global Environmental Change*, *16*(3), 253–267.

Gidengil, E. and Stolle, D. (2009). The role of social networks in immigrant women's political incorporation. *International Migration Review*, *43*(4), 727–763.

GoK. (2010). *Agricultural Sector Development Strategy 2010–2020*. Nairobi, Kenya: Government of Kenya.

Gunderson, L. H. (2001). *Panarchy: Understanding Transformations in Human and Natural Systems*. Washington, DC: Island Press.

Hickey, G.M., Pelletier, B., Brownhill, L., Kamau, G.M., and Maina, I.N. (2012). Preface: Challenges and opportunities for enhancing food security in Kenya. *Food Security*, *4*(3), 333–340.

Holling, C. S. (1973). Resilience and stability of ecological systems. *Annual Review of Ecology and Systematics*, 1–23.

KFSSG. (2013). *The 2013 Short Rains Season Assessment Report*. Nairobi, Kenya: Government of Kenya.

KFSSG. (2014). *The 2014 Long Rains Season Assessment Report*. Nairobi, Kenya: Government of Kenya.

Lowitt, K., Hickey, G.M., Ganpat, W., and Phillip, L. (2015). Linking communities of practice with value chain development in smallholder farming systems. *World Development* 74: 363–373.

Mayunga, J. S. (2007). Understanding and applying the concept of community disaster resilience: A capital-based approach. *Summer Academy for Social Vulnerability and Resilience Building*, 1–16.

Miruka, M., Okello, J., Kirigua, V., and Murithi, F. (2012). The role of the Kenya Agricultural Research Institute (KARI) in the attainment of household food security in Kenya: A policy and organizational review. *Food Security*, *4*(3), 341–354.

Nelson, V. and Stathers, T. (2009). Resilience, power, culture, and climate: A case study from semi-arid Tanzania, and new research directions. *Gender & Development*, *17*(1), 81–94.

Newman, L. and Dale, A. (2005). Network structure, diversity, and proactive resilience building: A response to Tompkins and Adger. *Ecology and Society*, *10*(1), r2.

Norris, F. H., Stevens, S. P., Pfefferbaum, B., Wyche, K. F., and Pfefferbaum, R. L. (2008). Community resilience as a metaphor, theory, set of capacities, and strategy for disaster readiness. *American Journal of Community Psychology*, *41*(1–2), 127–150.

Pretty, J. N. (1995). Participatory learning for sustainable agriculture. *World Development*, *23*(8), 1247–1263.

Putnam, R. D. (2001). *Bowling Alone: The Collapse and Revival of American Community*. New York, NY: Simon and Schuster.

Reed, M. S. (2008). Stakeholder participation for environmental management: A literature review. *Biological Conservation*, *141*(10), 2417–2431.

Rossi, A. and Lambrou, Y. (2008). *Gender and Equity Issues in Liquid Biofuels Production: Minimizing the Risks to Maximize the Opportunities.* Rome, Italy: FAO.

Rubin, A. and Babbie, E. (2007). *Research Methods for Social Work.* Belmont, CA: Brooks/Cole, Cengage Learning.

Swanstrom, T. (2008). Regional resilience: A critical examination of the ecological framework. *Institute of Urban & Regional Development.*

Tiffen, M., Mortimore, M., and Gichuki, F. (1994). *More People Less Erosion: Environmental Recovery in Kenya.* Chichester, UK: John Wiley & Sons.

Tompkins, E. L. and Adger, W. (2004). Does adaptive management of natural resources enhance resilience to climate change? *Ecology and Society, 9*(2), 10.

Tuana, N. and Cuomo, C. J. (2014). Climate change—Editors' introduction. *Hypatia, 29*(3), 533–540.

Turner II, B. L. (2010). Vulnerability and resilience: Coalescing or paralleling approaches for sustainability science? *Global Environmental Change, 20*(4), 570–576.

4 Land to feed my grandchildren

Grandmothers' challenges in accessing land resources in semi-arid Kenya

June Y. T. Po and Zipporah Bukania

Introduction: Women and access to land resources in sub-Saharan Africa

In the rural areas of many developing countries, women face more barriers than men in accessing productive factors, especially land (Agarwal, 1988, 1994a, 1994b; Besteman, 1995; Carney, 1988; Carney and Watts, 1991; Gray and Kevane, 1999; Koopman, 2009). This prevails despite the fact that over 70% of smallholder farmers in sub-Saharan Africa are women (Alliance for a Green Revolution in Africa [AGRA], 2014). In many Kenyan communities, women are customarily entitled to usufruct rights to cultivate the land of their husbands or fathers. Such secondary access to land resources through kinship persists today, and remains tenuous in many patrilineal and even in matrilineal inheritance systems. In many cases, widows, divorcées, and women who are in informal unions or have not married lose access to their land, making them and their children more vulnerable to food insufficiency and malnutrition (Gray and Kevane, 1999). Studies on women's land rights have shown empirically that these critical entitlements are correlated with increased empowerment and better outcomes for women and children (Agarwal, 1994a, 1994b; Doss, Kovarik, Peterman, Quisumbing, and van de Bold, 2013; Frankenberger and Coyle, 1993).

Traditional means of accessing land resources in Kenya are shifting. Demographic trends—such as a generally ageing population, increased attendance of girls in school, single motherhood, decreased rates of marriage, and, particularly, grandparents nurturing grandchildren in multigenerational or skipped-generational households—highlight the increasing importance of women's entitlements to land resources. Changes in formal and informal institutional processes, such as land privatisation and customary inheritance, likewise have overarching impacts on women's access to land resources (Food and Agriculture Organization of the United Nations [FAO], 2011a). A pertinent example, in our case, is the major reform in land inheritance introduced in Kenya's 2010 constitution (Republic of Kenya, 2010), further described in this chapter.

Our aim in this chapter is not to survey the literature on Kenyan women farmers and their entitlement to land for cultivation. Instead, we seek to shed greater light on the situations surrounding grandmothers' entitlement to land for nurturing their grandchildren—an illustrative case of one group of marginalised caregivers. In Africa, there are an estimated 38 million elderly people above 60 years old. This age group is projected to reach between 203 and 212 million by 2050 (HelpAge International, 2002; Lekalakala-Mokgele, 2011). Yet, very little is known about how grandparent-maintained households fare with respect to food security. Due to their declining health, and the growing trends of grandparents rearing grandchildren in rural communities, it is becoming critical to understand their challenges and strategies for food and nutritional security. Recognising this need, this chapter discusses some of the challenges that grandmothers face in terms of security of land tenure, which weaken their efforts to provide nutritious food for their grandchildren within the semi-arid drylands of eastern Kenya. We specifically focus on grandmothers in semi-arid villages in Makueni County.

The first part of the chapter briefly introduces existing formal policies as well as local norms and values related to women's entitlement to land resources within the Kamba ethnic community in the former Eastern Province of Kenya. In the second part, we draw out the changes in land entitlement for grandmothers in relation to men in the family and the attitudes grandparents have towards the growing trend of rearing grandchildren. Thirdly, we compare two cases to contrast the dynamics of land access in typical rural scenarios and illustrate the multiple layers of constraints that grandmothers face in caring for their grandchildren. Finally, we discuss the various livelihood assets and strategies grandmothers engage in to enhance their social and ecological resilience, in the face of old age, as well as in the maintenance of access to food, dietary diversity, and the nutritional security of grandchildren and other household members.

Land and land policies in Ukambani

To place the specific challenges of women's land resource management within the context of the semi-arid regions of Kenya, we first briefly describe Kambaland, or Ukambani. This is where the Kamba people, the fifth largest ethnic group in Kenya, settled in the semi-arid eastern region south and east of Mount Kenya more than 200 years ago. Farmers cultivate small parcels of land, majority of them less than two hectares, often with low yields due to the arid and semi-arid nature of the region. Annual average rainfall is less than 90 mm (Jaetzold, Schmidt, Hornetz, and Shisanya, 2006; Kaplan, 1984). Few farms have access to irrigation and most farmers practice low-input agriculture, depending on manure from livestock for soil fertilisers. In agro-pastoral regions, staple crops include maize, sorghum, millet, and legumes such as green grams, dolichos lablab, and pigeon peas; cattle, sheep, goats, and poultry are also reared. In the more arid regions of Ukambani, household livelihoods depend centrally on pastoralism since crop failure is relatively high.

Land in Kenya is generally categorised into freeholding land, government or trust land, and common land. Freeholding lands are individually owned, either inherited or purchased. Inherited land from ancestral land can be described as plots of land where people initially settled that was not occupied by another person. They gained ownership of the land through input of labour, such as turning the uncultivated land, *kitheka*, into farms, *shambas*, or by building a dwelling, *boma*, on the land.

In Kenya, this was the norm when land was abundant before British colonial times. Currently, most smallholder farmers in upper Makueni County are cultivating on free-holdings. But other forms of tenure co-exist. A male household head recounted to us how his mother settled in the area: "We came here just to have a share to get somewhere to dwell because nobody own[ed] this area… [the land] was not for anybody, it was a public area."

The second major category of land is government or trust land. Here, the government retains the land rights. Considerable land in Ukambani, especially in Makueni County, was allocated to farmers through land settlement schemes. For example, inhabitants who have been displaced from lands that have been previously set aside for national game reserves and parks, now cultivate on trust lands given by the government for resettlement. Settlement schemes within Kambaland originated from the colonial era. Kamba men who had served in the British army overseas during the Second World War were among those originally settled by the colonial government in the post-war period. There was also a soldier-settlement scheme in Kenya for Europeans, but the landplots offered to them were of higher altitude and greater agricultural potential. Further sub-division of the lands started in the 1960s in upper Makueni County, in areas like Wote, closer to Nairobi. Many farmers either had government surveyors adjudicate their land and received a land plot number or they registered for a title deed. However, in lower Makueni, the process of land titling started later. For example, in Kibwezi, the adjudication process started in the 1990s, and the majority of farmers do not yet possess title deeds twenty-five years later.

Thirdly, common lands are used, managed, and protected by the community. They include sites such as school grounds, forests, common paths between farms, springs, and sometimes dams, that are built by the community with financial support from non-governmental organisations (NGOs) or government programmes. Under the three categories of land holdings, this chapter focuses on the nuanced gendered access to land resources in freeholding lands, which are tightly intertwined within institutions of marriage, family structure, and inheritance.

In 2010, Article 60(1) of Kenya's newly promulgated constitution announced "the elimination of gender discrimination in law, customs and practices related to land and property in land." This clause introduced the formal rights of women to inherit land, addressing a common form of discrimination against unmarried, widowed, and divorced women (FAO, 2007). The constitutional change was publicised in the media as an era of new rights for daughters to inherit land from their fathers. Three new land bills were passed in 2012, repealing existing land laws, and signalling movement towards enacting constitutional mandates. The views and reactions of Kamba villagers to these changes are shared in this chapter.

Women's customary land entitlements in Ukambani

The Kamba ethnic group have historically practised both polygamous and monogamous marriage within a generally patrilineal society (Tiffen, Mortimore, and

Gichuki, 1994). Many Kambas maintain associations with their clans, although clan cohesion is gradually weakening. Some aspects of the traditional beliefs in magic and family curses that were common at the turn of the 20th century (Hobley, 1910) appear to remain a part of Kamba society; interestingly, often associated with family land disputes. However, the predominant religious belief in the Kamba communities is Christianity and Catholicism. In Ukambani, as elsewhere in Africa, high fertility rates and single motherhood, as well as the HIV/AIDS pandemic that began in the 1980s, have resulted in the emergence of skipped-generational family structures wherein grandmothers become the primary caregivers for their grandchildren (Linsk and Mason, 2004; Nyambedha, Wandibba, and Aagaard-Hansen, 2003; Omariba, 2006). This phenomenon gained attention globally through movements such as Canada's Stephen Lewis Foundation, which raises money for grandmothers who raise AIDS orphans. By exploring the grandmothers' access to land resources and livelihood strategies in semi-arid Kenya, this chapter may lead to a better understanding of the similar fates of grandmother caregivers elsewhere in Africa. In order to illustrate the structural challenges grandmothers face in accessing land resources and providing nutritional security for their grandchildren, it is beneficial for readers to have a brief understanding of land resource entitlement by women within their relationships, as wives, mothers, daughters, and widows, within the Kamba culture.

According to Kamba traditions, as explained to us by women and men in Makueni, women have customary entitlement to land for cultivation from their spouses, but women are not entitled to inherit ancestral lands from their fathers. In order to access land, one of their major livelihood resources, women are expected to become a wife. "Whoever marries has a piece of land," a man explained during a men's-only focus group discussion. Not only are women expected to be married, men are also expected to display signs of maturity, often through starting a family, before they are given their own piece of land by their fathers to build a home and cultivate on a farm independently. "The husband had pieces of land so [the wife] possess[es] the land together with the husband." Wives hold customary ownership of their spouse's land after the spouse dies, but young widows often face challenges different from older widows who have established their relationships and status within the household as grandmothers.

In the Kamba tradition, when a woman marries into the man's family, the groom's family pays the dowry, or bride price, which includes a minimum of three goats. After marriage, the wife no longer belongs to her father's clan. She belongs to the husband's clan. She will rely on her husband and his parents for her livelihood. This separation of women from their natal clans is one of the primary justifications for women losing entitlement to inheritance of ancestral land in their natal village. If a wife formally separates from her husband, traditional processes are well understood and in place to reconcile or recognise the separation. If the wife initiates the separation from her husband, and her natal family formally pays the husband's family back one goat, as the "refusal goat" (*mbui ya ulee*); this act signifies a complete separation from the husband's family and clan. She relinquishes her customary entitlement to any kind of land or property from the husband's family. One of the significant implications of this is that when the wife dies, she will not be buried in her husband's ancestral land. Yet if the refusal goat was not paid, the husband is entitled to have his wife's body brought back to be buried in his land even if they have separated. After separation, a wife has an option to return to her natal family; however, divorced women often remain ostracised by their families and natal clan, in part due to reluctance to provide her with land.

If the wife is forced to leave the husband's family or is "chased away" without any customary rites performed, her bond to the husband's family and clan remains intact. In other words, the husband's family's obligation to the woman's entitlements remains intact as well; though having been chased away, her ability to realise her entitlements is severely limited. As the mother of her husband's children, she remains entitled to be buried in the land of her husband. The children are entitled to land from their father when they mature and start their own families.

According to customary norms, young widows should continue to farm on their husband's land. They have increased decision-making power over agricultural practices, but they remain under the authority of their in-laws when it comes to making decisions over large sales and purchases. We were told of numerous cases where relatives would intervene after the husband passed away to make decisions on land resources and assets on behalf of the husband, especially if the father-in-law or mother-in-law was not present. In scenarios where the husband's land is coveted, the widow can be blamed for a number of faults to justify casting her out of the husband's family. She could even be blamed for the husband's death, or harassed by neighbouring siblings-in-law and other relatives until she is coerced into giving up her rights to her husband's land. The relatives may also rule in favour of giving the children the land instead of transferring formal ownership to the widow. Moreover, the relatives generally rule traditionally in favour of the sons and not the daughters of the widow.

Some customary protections for women are in place, but not often accounted for in planning. For instance, one man explained, "[I]f you have sons and daughters, [a daughter] who doesn't get married will have the right to be given a portion of that land." In contrast, in most cases, the return of a daughter from a failed marriage is an unplanned event which the family does not account for, hence the family's head does not allocate land resources for her. This is frequently justified by the limitations of land already allocated to sons in the family who require the parcels of land for their own families. If the land has been allocated to the sons in the family for cultivation, the parents traditionally leave a piece of land for themselves where they cultivate with the last-born son. Once the parents pass away, the last-born son inherits that piece of land in addition to any piece of land that the parents previously allocated to him. In other cases, the eldest son inherits the land from the father and is instructed to subdivide the land among the siblings, brothers, and unmarried sisters before the father passes away. When a daughter returns from a marriage separation, it poses an economic shock to the family system in increased household dependents' expenditure and land resource supply. Although the daughters are entitled to a sub-section of the land remaining for the parents, they are often refused by their family members. "Wise parents would leave a piece of land for themselves in case of daughters who come back," a mother said in an interview in Makueni County.

The land inheritance clause in Kenya's 2010 constitution does not specify conditions of marital status of the daughters for their rights to claim land from their father. Most respondents and research participants expressed alignment with the Kamba traditional institutions. The majority of them interpreted the constitutional clause as: if a daughter remains unmarried, she is entitled to a piece of her father's land to cultivate and support her children. They maintained the prevailing value—that married daughters should rely on their husband for land resources and are not entitled to land inheritance from their father. One respondent stated, "When she is married she can't be given a piece of land. You should stay there and use the property of your husband." The justification is based on the perceived equity of "inherited" land holdings between man and woman. Another respondent explained, "She cannot own two lands, [from] her husband and [from] my *shamba*. No, she has to have one *shamba*." (See Box 4.1.)

Changes in access to land resources for grandmothers

In comparison to the insecurity of secondary access to land resources that many women experience, a grandmother may have increased access to the land. Within the family, a grandmother maintains her role as a wife, a mother and a mother-in-law as well as a grandparent. If her husbands' land has been subdivided to her sons, she is less likely to be dispossessed by relatives from the land she is cultivating. If her husband is not alive, customarily, she becomes the head of the household. Within the community she has established her social networks, relationships, and status in the period of time since her marriage. Having lived in the community for many years, the grandmother may have established extensive relationships within the

---

Box 4.1 Women purchasing, leasing, and sharing land

Besides accessing land through marriage and kinship, the last decade has seen a rise in women purchasing land on their own, in groups, and having their name on title deeds, either in co-ownership or individual ownership. In Makueni, we met women who contributed money to purchase land with their husbands, where the name on the title deed or land number, a precursor document of the title deed issued by the surveyor, only had the husband's name. In other instances, women hire or lease land by the season, either individually or as an organized women's group or as a mixed, male and female farmers group.

As an example of group dynamics concerning shared use of land, one participating farmers from the KARI-KEMRI-McGill food security project steped forward to allocate a piece of their own land for the establishment of a demonstration farm. The farmers group leadership committee and group members established written agreements concerning the use and sharing of the land, the work, and the outcomes of their efforts. Other groups leased land for the demonstration plots and similarly took formal group decisions about sharing of responsibilities and benefits.

community by participating in women's groups, clan and family meetings, helping organise multiple religious celebrations, engagements, marriage negotiations, weddings, funerals, and attending innumerable types of fundraisers of other families. The social networks, reciprocity, and trust she has established in her community not only add to her social capital, but, often, her status through time. In turn, the grandmother uses her social capital and social status to better secure her access to land resources as the major physical capital for her livelihood. In the last section of this chapter, we use the sustainable livelihoods framework (Chambers, 1995; Ellis, 2000; Scoones, 2009) to synthesise the potential strategies grandmothers have at their disposal to enhance their resilience to shocks in old age and in securing their access to land resources.

There are certain familial norms that protect a grandmother's access to land resources in our study area. For example, her adult children must all agree informally through discussions or formally in court upon decisions surrounding land division and sales. With their grown children's assistance, the family may decide to transfer the title deed from the name of the deceased father to their mother's name. Yet, with a relatively secure access to their land, grandmothers face other kinds of difficulties related to a rising trend of nurturing their grandchildren as primary caregivers.

Growing trends of grandparents rearing grandchildren

Grandchildren are left in the care of their grandparents either because the children's parents have died, or they have left to look for waged work in urban centres. "For the first month or two, they would send money home. After that, they disappear and we do not know where they are," one grandmother lamented. If caring for the number of children in the family is beyond the abilities of the parents, grandparents often step in. This is seen as 'sharing the load.' When the parents are not around, most grandparents cannot abandon the young children if they have the ability to support them. "We see [the grandchildren] as our own children. We are responsible for them."

Taking on the responsibility of primary caregiver to grandchildren is not widely perceived as having a negative impact on prior household consumption and livelihood stability; in contrast, it is more often seen as a common familial obligation (P. Wambua, personal communication, August 3, 2013). This is more apparent in rural communities where men may only participate in the upbringing of children financially, but not physically on daily care and feeding. Due to changing cultural and social circumstances (see Box 4.2), gendered responsibilities and roles in our study area are beginning to change, with more men increasingly participating in childcare.

In many cases, both grandparents take part in raising funds for school fees. If the grandparents are able, they support their grandchildren through secondary and post-secondary education. Grandparents stated that one of the major challenges of rearing grandchildren is paying school fees. The grandfathers and grandmothers echoed each other in their experience of coping with raising their grandchildren. It is

Box 4.2 Factors contributing to skipped-generational households

Some of the factors contributing to the growing prevalence of skipped-generational households are increased cost of living, increased numbers of single mothers, and orphaned children due to the HIV/AIDS epidemic. Among those interviewed, many attributed increased cost of living as the reason young men hesitate to marry and start a family. Although economic barriers hinder formal marriages, whether customary, religious, or civil, fertility rates remain high and contraceptive use remains low. Having children out of wedlock is very common.

Literature across developing countries provides strong evidence that additional years of girls' education are correlated strongly with delayed childbearing (Lloyd and Mensch, 1999). Yet within the rural Kamba regions, prevalence of young single mothers remains relatively high. They may not get support from the children's father, and are often left with limited options of rearing the children on their own or leaving them with their grandparents while the mothers search for casual employment in urban centers. Given the small salary young women may earn in the urban centers, grandparents expressed low expectation of having their daughters send any money home to support their children.

Finally, there is a high prevalence of HIV/AIDS in Ukambani, affecting both women and men. Orphaned children are most often left under the care of their grandparents. There are local NGOs, such as the Orphaned and Vulnerable Children, which assist orphans and the elderly affected by HIV/AIDS deaths. Such organizations support the children with school expenses, but rarely do they allocate funding to nutritional needs of the elderly grandparents and their orphaned grandchildren.

common to involve communities in public efforts to raise school fees through community fundraisers, better known as *harambees* (Kiswahili for "Let's all pull together").

Grandmothers generally have the gender-defined responsibilities of providing food and clothing and passing on stories of Kamba traditions, customs, and morals, as well as skills around the kitchen and the virtues of sanitation and hygiene. Grandfathers are generally more involved with transferring agricultural and trade skills, such as maintenance of the plough, keeping of the cattle and goats, and sometimes poultry. In cases where the grandfather is not working and is not contributing funds to support the grandchildren, either due to poor health, lack of employment or costly habitual expenses like alcohol and cigarettes, the grandmother cope by searching for employment on neighbouring farms as a casual labourer or by selling artisanal crafts such as hand-braided ropes or baskets made from sisal fibres, often produced in periods off the farm, or while walking from field to field.

Land to nourish my grandchildren

Grandmothers' responsibility for feeding the grandchildren places them in a specific relationship with the grandchildren. Not only does she shift her retirement plans, along with her division of harvest for domestic use or market sales, but she may also start cooking again for her grandchildren. In a multigenerational household, the grandmother's daughter-in-law usually cooks and serves her. She is the wife of the youngest son, who traditionally lives with her mother-in-law. In the literature, there have been cases of surrogate motherhood where grandmothers breastfeed young infants in order to sustain the lives of their grandchildren who have lost their mothers or who may be infected with HIV mother-to-child transmission. Partly, this was tried as a last resort by the grandmothers, who did not have enough money to buy milk formula or cow's milk for the grandchildren (Oguta, Omwega, and Sehmi, 2004). In more prosperous cases, grandmothers can indirectly benefit from the land through her relatives, and other sons and daughters. The grandchildren might have uncles and aunts to help the grandmother, contributing food, money or both. They may contribute part of the harvests from their own farms to their mothers, or they may harvest together in an extended compound, even storing the harvests in the same granary. However, grandmothers explained that these are rare occurrences as their sons have their own families to support, their own mouths to feed. If they do bring money or food to their mother, one grandmother explained, she would welcome it as well.

Besides kinship networks, grandmothers who are able to participate in community groups, such as church committees, women's microlending groups, revolving self-help groups or farmers groups have additional means to transform their social networks, participation, and reciprocal trust into resilient mechanisms to support their grandchildren. One of these mechanisms is collective harvesting to grow more food for their households. Many grandmothers in our study reported being involved in farmers groups to share the labour burden of managing a farm. Yet, many grandmothers encountered physical or economic barriers to benefitting from their social surroundings. For example, some grandmothers have withdrawn from farmers groups that primarily dig terraces together in each member's farm. The grandmothers had physical ailments which prevented them from providing their share of labour to the group. Others chose not to join a revolving savings group because they could not afford the amount each member was required to contribute at each meeting, limiting their access to credit and loans to support their livelihoods. In more strenuous cases, grandparents who own land may use their title deeds, if they have one, as collateral to borrow money from formal banks. This risks having their land taken away by the banks if they are unable to repay the loan and interest. In other cases where the family does not have official documents to apply for a loan, they may even use the corrugated metal sheets on their roof or other physical materials of value as collateral. This demonstrates the dire circumstances some find themselves in, and the consequences of default (removal of the home's roof) contributes greatly to the reluctance of home owners to apply for formal loans as a means to advance their livelihood strategies (more details on financial lives of smallholder women in Chapter 6, this volume).

As described above, grandparents go through additional stresses to provide food for their grandchildren. Yet, grandparents' apparent low nutritional awareness and some unhealthy food habits pose further concerns for the grandchildren's nutritional state. Basic available foods in eastern Kenya are highly cereal-based, predominantly corn, rice, wheat, sometimes legume, and lentils. Grandmothers' food and nutrition knowledge impacts the dietary diversity they are able to provide for their grandchildren. There are age-specific nutritional requirements that demand different types of food as the child continues to grow: carbohydrates for energy, proteins for growth, micro-nutrients for critical functioning of the body's metabolism and cognitive development. Younger children, especially under the age of five, are most vulnerable and susceptible to infections, therefore they have increased needs for protective foods in addition to energy and body-building foods. Addressing age-specific dietary needs sometimes proves challenging due to environmental constraints and a lack of availability of affordable nutritious foods.

Sometimes highly nutritious foods are available, but they are not incorporated into the diet. For example, a woman might sell eggs and use the income to purchase sugar or bread even though eggs have more nutritional value. Pumpkin and arrowroots are commonly grown in the farm and eaten during breakfast, but they take more firewood to cook and generally take longer to prepare, making them less convenient than white bread and margarine. Local indigenous leafy vegetables and wild edible plants are high in minerals and vitamins, but are often not cooked for family meals due to their association with poverty, crop failure, and low social status. Indigenous crops that are high in nutrients and grow abundantly in the semi-arid regions, such as sorghum and pearl millet, have lost favour to corn, rice, and wheat due to various reasons, mainly rooted in colonial land and labour policies (Brownhill, 2009).

Higher value is perceived in processed, convenient food bought with money in the market, such as sweetened white bread over farm grown produce, such as cassava. Moreover, processed food such as potato chips, candies, and soda beverages that have high sugar, salt, and fat content are welcomed by both young and old in the rural communities. While mothers or grandmothers may buy nutritious food from the market, as urban dwellers do, instead of consuming the products of their land, the farmers interviewed also expressed the vulnerability to food price volatility and the added security they have of providing enough food from their own farms.

Comparing the cases of Anna and Lilian

To illustrate the multiple layers of constraints that grandmothers face in caring for their grandchildren, we compare two cases to contrast how land is accessed by women in typical rural scenarios in Makueni County.

The case of Anna: "The land is not mine to give to my daughter"

Anna (pseudonym) lived on a farm in Mumbuni sub-location. Her husband stayed at home due to old age. As the primary farmer of the land, Anna prepared the land before each rainy season with a hoe, ox, and plough to cut the ridges and to drop the seeds. On her husband's approximately one-and-a-half-acre plot of land, she planted maize, pigeon peas, sorghum, and cow peas for subsistence, though she recalled there was no harvest for cow peas in the past season. Everything she ate was grown from her farm except for kale, which she bought from the market. She got milk from her two cows. Before the long rainy season when she experienced a shortage of food, she performed casual labour on other people's farm for cash to buy food from the market, but chances for employment were far and few between.

Besides the piece of farm for her and her husband, she used remittances from her unmarried daughter to hire a piece of land from her sister-in-law. The rent was 1,000 Kenyan shillings (KSh) (approximately $11 USD) per season. On that land parcel, she grew cotton and maize to support her three grandchildren aged 7, 9, and 10 years. Last season, she sold a total 2,000 shillings of cotton and used the earnings to buy seeds for food crops in that planting season. In recounting the reason for hiring a piece of land, she said it was to avoid having to use the ancestral land, which had been informally divided to her three sons. Anna's sons were allocated the piece of land to cultivate approximately ten years ago, when they got married. Anna explained that it was difficult to stay with her sons' families as division of labour became problematic. "Someone may not want to cook or fetch water and conflicts may arise." The sons who were cultivating their pieces of land do not hold title to the land. The father of Anna's husband was the one whose name was on the title deed, although he was no longer alive. He had had two wives. Until the sons of the two families agree with each other, the process of acquiring a title deed for Anna's husband will not begin.

Anna's daughter, on the other hand, was never married and had given birth while staying with Anna. "Yeah, I told her not to abort, [but to] just give birth and bring [her children] to me," said Anna. Since her daughter was unmarried, according to Kamba tradition, she was entitled to her father's land for cultivation to support her children. "She has not been given…the *shamba* is for the grandfather, the father of my husband." When asked, "But [do] the sons have some piece of land from your husband?" Anna responded, "My children? Yes, that *shamba*." When asked, "Can your daughter also be shown a piece of land to cultivate?" Anna replied, "If it could be mine, I can give [it to] her…But the land is not [mine]." She explained, "She could not, [it would] bring much tension … [I]f you bring tension [by asking for land] you can be evacuated with your children. Where will you go?"

As a married woman, Anna had traditional land entitlement but since her husband was still living, decisions concerning land were in his hands. She explained, "Furthermore, it is not mine. Even [if] she want[ed], where will I get [the land] from?" She cannot ask her husband, "because the *shamba* is not for [my] husband and we are many," indicating there were many brothers of the husband. "[W]hen you are my brother, when you show your daughter a piece of land, and even you, yourself, you are not entitled to that piece of land… [B]rothers can raise many questions." When asked whether Anna's sons could allocate a piece of land to their sister, she revealed, "Now the ones with wives cannot accept [the request] to cut [a piece of land] where they [have been] cultivat[ing]." Her three sons could allocate a piece of farm to their sister: "[|T]hey don't have a problem with that, but their wives…they can't move a piece for her, even a small one." If husbands allocate land to their sisters, they are, in effect, taking land away from their own wives. This makes wives reluctant to agree for men to share land with their sisters. We were told, "[I]n two to three days, they [wives] start conflicts. There will be conflicts every day."

On the hired land, Anna was allowed only to plant food crops, not permanent crops such as fruit trees. She must seek permission for each agricultural decision from the wife of her brother-in-law, who was the owner of the land after the brother-in-law had passed away. Not only did each decision require ongoing negotiation with the sister-in-law, production was limited and short-term. This constrained access to land resources and undermined household food security, as women were not free to plant what and when they want, and may lose access altogether, especially if the owner ceases to rent out the land. "If I had one [piece of land of my own], I can cultivate and at long last, I [will] leave it to [the grandchildren]." We followed up: "If you would have your own land, can you give it to your daughter?" Anna answered, "*Kabisa* (Straight away)."

The case of Lilian

Lilian (pseudonym) was born around 1930. She lived in a grass thatched mud hut on the top of a rocky hillside. The land ran steeply down towards the sandy dried bed of the river, where they fetched water each day. She was married to a woman when she was very young.

The female-husband (see Box 4.3) provided for Lilian, allocating a portion of land for her to cultivate and feed her children. By the time of our interview, both her female-husband and the original male husband had passed away. Lilian now owns the land. Although she had full customary entitlement and ownership of the land, she was struggling with cultivating enough food for her grandchildren. In addition, she wove sisal ropes and sought casual labour almost every day. The day

---

Box 4.3 The female husband

In Kamba tradition, a married woman who is barren or cannot produce a son may marry a younger woman as her 'wife.' The barren woman will then take on the role of the "female-husband" of the young woman. All rituals of these marriages are observed, dowry, or bride-price, is paid to the young woman's father, and all rules of divorce in the society apply (Herskovits, 1937, p. 335). In this circumstance, the marriage differs from lesbian marriages as the union is legally and socially accepted but the relationship is not sexual. The expectation for the young woman is to provide a male heir for the female-husband. She likewise expects to be housed and given access to land and other livelihood resources. The female-husband takes responsibility for, and has rights over, any children born by the bride. The biological father is someone outside the marriage, and he does not have rights over the children or any of the resources of the marriage. Through a woman-to-woman marriage, the female-husband protects her access to material wealth, especially land resources and family inheritance, by acquiring a wife and producing son who perpetuates the family name.

we visited, there were two young girls playing around the homestead. Her grandson was a secondary school student. He had become involved with some troubles and was summoned to court. In order to pay court fees, Lilian sold portion after portion of her land. Ultimately, she ended up selling the piece of land where her thatched mud hut stood. The only reason she said she was still staying there was because she lacked funds to move and rebuild in another location. The buyer sympathised and allowed her to stay on the piece of land until she could gather the funds to relocate. She pointed to the tiny shape of a homestead across the river-carved valley and told us that, after the interview, she would trek across the valley to her neighbour and borrow some food to cook for the evening's meal.

Five months later, when we visited the widows group meeting again and met Lilian, her grandson's case was ongoing, but she had relocated to another piece of land. She said that things were better since the relocation. The owner who had previously bought her land needed to start using the land. So the owner lent her some money to relocate, which she will pay back through installments. She has since rebuilt her home even though she continues to struggle for her livelihood.

Grandmothers' sustainable livelihoods, assets, and strategies

From a grandmother's perspective, a woman's secure access to land holdings can have a tremendous impact on her and her household's food security. In Anna's case, the local institutional processes did not provide her with the secured resources she needed to support her grandchildren. She had access to financial capital to hire a piece of land and access to labour by hiring casual workers, such that she could benefit from cotton sales. This relatively fortunate case depicts circumstances where financial capital is present but the main constraint is natural capital: available land for cultivation. Anna expressed that she hired a piece of land to avoid cultivating her sons' farm in order to keep peace within her family. Even though there seems to be a traditional hierarchy within the family whereby the wife submits to the husband in decision making and the sons and daughters obey their parents, this case highlights that the reality is often far more complex. Anna employed a livelihoods strategy which forwent entitlement to land and prioritised harmonious family relationships. Anna's daughter was unmarried with three young school-age children. According to Kamba traditions, she was entitled to a piece of land from her father for her and her children to grow food. Yet, often patriarchal norms and customary institutional processes described in a best-case scenario fall short in actually protecting marginalised members in the household and the community.

In Lilian's case, she had full ownership of her land, albeit small, rocky, and with low productivity. Although she eventually had to sell the land upon which her thatched dwelling was built, she was able to generate funds when needed to help her grandson in judicial court fees. Both cases portrayed grandmothers and grandchildren embedded in structural poverty. Lilian, who was significantly poorer than Anna, was able to use her land as a physical asset, and her entitlement to sell, as her last resort to care for her grandchild. Anna, who had financial capital, acquired land through lease-agreement, but she could not solely make decisions on the land she cultivated, and in turn also struggled to ensure food security for her grandchildren. Which case was better? Does grandmothers' ownership of land empower them to seek alternative livelihoods strategies? Would owning land resources lift them out of structural poverty?

In general, we find grandmothers, as well as other smallholder farmers, have poor access to irrigation and other mechanised farming technologies and poor access to farm inputs, market, credit, and unemployment. As more youths seek jobs in towns, farm labour is also diminishing within rural communities. With constrained access to information and knowledge, farmers often succumb to greater exploitation from 'middlemen' during the sale of harvests. However, using the sustainable livelihoods framework (Chambers, 1995; De Haan and Zoomers, 2005; Ellis, 2000; Scoones, 2009), we find that grandmothers have access to specific livelihood assets that enable them to be quite resilient members of the community. Here, we look more closely into the diverse natural, physical, human, financial, and social capital they have as their livelihood assets.

Considering natural capital, smallholder farmers in our study area generally have lower access to water resources; many hours and much energy are spent on sourcing water from sand dams, rivers, and groundwater pumps. Yet, a grandmother's relation with her husband results in greater opportunities to benefit from land as a natural capital even without formal ownership. She may not be able to sell the land, or acquire loans using the title deed as collateral, but her years of experience in the community enable her to better control and secure her access to farmland. Although her physical capacity to engage in manual labour is reduced, she has family members, relatives, neighbours, and other community group members who may share labour in times of tilling, weeding, or harvesting, enabling them to increase their crop yield and benefits from the land. Other natural capital that grandmothers and women have specific access to is poultry. Compared to other livestock such as goats, sheep and cattle, poultry-rearing is predominately within women's domain (see Chapter 7, this volume). In addition, grandmothers often have better knowledge of the surrounding environment. They are familiar with locations of available wild edible foods and fibres, which they may collect themselves or instruct family members to forage to booster food and nutrition security for the family (Shumsky, Hickey, Pelletier, and Johns, 2014).

This unique knowledge of wild edible plants, their locality, seasonality, and nutritional and medicinal benefits is an often undervalued form of human capital, which is unlikely acquired from a book or from a classroom. The authors recognise the incredible value gained from indigenous knowledge that is passed down across generations, knowledge that is slowly disappearing. Like the local Kikamba dialect, which is only taught until early levels of primary education, knowledge of local language and customs is preserved through storytelling. One grandmother told the authors it is her responsibility to transfer such knowledge to her grandchildren. Although many grandmothers we interviewed in rural and impoverished communities had little or no formal education, they had a wealth of life experience in Kambaland: having survived through major political upheavals and some of the most serious historical droughts in the area, among other harsh conditions, they have first-hand experience of what is required to be resilient in the face of shocks. For example, two grandmothers described life during the Mau Mau uprising before the independence of Kenya when they had to hide and live in the forests for months away from their homes.

Although caregiving poses additional stresses to grandparent-headed households, exacerbating their often compromised nutritional status, our observations suggest that grandmothers have a wealth of local knowledge and have developed effective coping capacities, having survived in resource-limited rural environments for decades—and that these capacities can better flourish when women have more secure access to land resources.

Grandmothers' physical and financial capital in the context of semi-arid farming systems are closely tied with her social capital. Many forms of physical capital that are measured at the national scale, such as infrastructure, telecommunication, and housing and farming equipment, are similarly important to the general smallholder farmer. Yet an increased wealth of social capital, accumulated through the years, provides grandmothers with some advantages over younger women in terms of accumulating access to resources. For example, older women typically have a greater range of mobility than younger women, because an elder woman travelling alone over long distances is less frowned upon by local society. They also tend to have more access through different communicative avenues to raise issues with chiefs, or government agencies, due to the longevity of their involvement in local politics that have in the past impinged directly on the everyday lives of local people. Young women who are newly married into the community are less capable of employing this form of social capital.

Grandparents who are not actively engaged in caregiving can have more time to engage in civic volunteering, such as helping with political campaigns within their local districts. This, in turn, increases their networks, status, and experience through community participation. We keep in mind that elders who are isolated due to reduced physical mobility and health will have varying, but generally falling degrees of access to social activities that take place outside the home. In such cases, the strong local practice of holding women's group meetings at the homes of members allows for otherwise isolated widows to share the fellowship, as well as the resources, spiritual support, and livelihood strategies that such social capital permits.

Nevertheless, our results indicated that grandmothers tend to gain more memberships in different groups, possibly including multiple microlending, table banking or revolving credit groups that meet once or twice a month. They were also more likely to be listed on the household rosters at the chief and sub-chief's office, enabling higher enrolment in farmers groups or development initiatives funded by NGOs as well as government and foreign international development agencies. Having more experience with decision-making and leadership within the community, grandmothers, as an age group, have rich knowledge of who are key actors and influential local leaders rules in the semi-arid community. This allows them to discern more successfully which efforts will be fruitful both in promoting individual and household assets and in building community cohesion.

Shaping policies with gendered resilience

By better understanding existing challenges that grandmothers face in accessing land resources, and their growing burden of rearing grandchildren, this chapter has aimed to open a discourse on how development policies, especially regarding agriculture and land, may be shaped within a more equitable gendered framework and context. Efforts in international development agencies have placed high importance on women's empowerment as the foundation to successful intervention and programmatic efforts. In a recent report by the FAO on "Women in Agriculture: Closing the Gender Gap for Development" (2011b), the importance and power of customary land rights is recognised, while noting the galvanising impact of educating and evaluating government officials along gender targets, raising women's legal literacy, ensuring that they understand their land, marital, and individual rights and ways they can be enforced and protected. Women organisations can also be effective in ensuring women's voices are heard. Rwanda provides one example of how state institutions and civil society organisations can better work together to reform land tenure legislation to strengthen gender equity (FAO, 2011b). Similarly, women's voices are given more attention through the participation of women in local government. For example, the constitution of Rwanda and Kenya mandates that 30% parliamentary representatives be women. In the United Republic of Tanzania, where local land disputes are settled by village land councils, three out of the seven members in the council must be women (Ikdahl, 2008). Although women's representation in natural resource governance have yet to reach such prevalence on the ground, we recognise that these mandates are a step forward in getting women's voices heard. Adjustments and assistance in bureaucratic procedures such as joint-titling have the potential to better protect women within marriage. Policies that support assistance in applying for land titles, such as acquiring official documents (e.g., birth certificates) or filling out forms can also narrow the gender gap (FAO, 2010). "Women must be an integral part of the implementation of the land programmes. Training community members as paralegals, topographers and conflict mediators can help build community skills that increase the probability that women's concerns will be addressed" (FAO, 2011b).

From our description of the specific livelihood assets of Anna and Lilian, and analysing the livelihood assets specific to grandmothers operating within the rural Makueni context, and local institutions and policies that may support or hinder their use of and benefit from assets, we broadly portrayed the livelihood strategies available to grandmother-headed households. Indeed, we found grandmothers to have greater amounts of physical, financial, and social capital than young widows or single mothers to help care for young children. Although offering examples of being resilient members of the community, the cases of Anna and Lilian show that their food and nutritional security is low and their vulnerability to internal and external shocks is extremely high. Ultimately, secure access to land resources offers an important way in which they can increase their social-ecological resilience (Adger, 2000; Bahadur, Ibrahim, and Tanner, 2010; Cote and Nightingale, 2012; Holling, 1973).

Such resilience can be supported through innovative land and agriculture-related policy frameworks that integrate mechanisms to foster gendered resilience within households, so that marginalised groups, such as single mothers and grandmothers, are supported to pursue livelihoods that offer food and nutrition security to themselves and their grandchildren.

References

Adger, W.N. (2000). Social and ecological resilience: Are they related? *Progress in Human Geography*, 24, 347–364.

AGRA. (2014). *Africa Agriculture Status Report 2014: Climate Change and Smallholder Agriculture in Sub-Saharan Africa*. Nairobi, Kenya: Alliance for a Green Revolution in Africa.

Agarwal, B. (1988). Who Sows—Who Reaps—Women and land rights in India. *Journal of Peasant Studies,* 15, 531–581.

Agarwal, B. (1994a). Gender and command over property—A critical gap in economic-analysis and policy in South-Asia. *World Development*, 22, 1455–1478.

Agarwal, B. (1994b). Gender, resistance and land—Interlinked struggles over resources and meanings in South-Asia. *Journal of Peasant Studies*, 22, 81–125.

Bahadur, A.V., Ibrahim, M., and Tanner T. (2010). *The Resilience Renaissance? Unpacking of Resilience for Tackling Climate Change and Disasters.* (Strengthening Climate Resilience Discussion Paper 1). Brighton: University of Sussex, Institute of Development Studies.

Besteman, C. (1995). Polygyny, womens land-tenure, and the mother-son partnership in southern Somalia. *Journal of Anthropological Research*, 51, 193–213.

Brownhill, L. (2009). *Land, Food, Freedom: Struggles for the Gendered Commons in Kenya, 1870–2007.* Trenton, NJ and Asmara: Africa World Press.

Carney, J. and Watts, M. (1991). Disciplining women—Rice, mechanization, and the evolution of Mandinka gender relations in Senegambia. *Signs: Journal of Women in Culture and Society*, 16, 651–681.

Carney, J. (1988). Struggles over land and crops in an irrigated rice scheme: The Gambia. In J. Davison (Ed.), *Agriculture, Women, and Land* (pp. 59–78). Boulder, CO: Westview Press.

Chambers, R. (1995). Poverty and livelihoods: Whose reality counts? *Environment and Urbanization*, 7, 173–204.

Cote, M. and Nightingale, A. (2012). Resilience thinking meets social theory: Situating social change in socio-ecological systems (SES) research. *Progress in Human Geography*, 36, 475.

De Haan, L. and Zoomers, A. (2005). Exploring the frontier of livelihoods research. *Development and Change*, 36, 27–47.

Doss, C., Kovarik, C., Peterman, A., Quisumbing, A.R., and van den Bold, M. (2013). *Gender Inequalities in Ownership and Control of Land in Africa* (Discussion Paper 01308). Washington, D.C.: International Food Policy Research Institute.

Ellis, F. (2000). *Rural Livelihoods and Diversity in Developing Countries*. Oxford: Oxford University Press.

FAO. (2007). *Gender and law. Women's Rights in Agriculture*. Rome, Italy: FAO.

FAO. (2010). *Land tenure, Land Policy, and Gender in Rural Areas*. Rome, Italy: FAO.

FAO. (2011a). *The Role of Women in Agriculture* (Agricultural Development Economics Division, ESA Working Paper). Rome, Italy: FAO.

FAO. (2011b). *The state of food and agriculture—Women in agriculture: Closing the gender gap in development*. Rome, Italy: FAO.

Frankenberger, T. and Coyle, P.E. (1993). Integrating household food security in Farming Systems Research/Extension. *Journal of Farming Systems Research/Extension*, 4(1), 35–65.

Gray, L. and Kevane, M. (1999). Diminished access, diverted exclusion: Women and land tenure in sub-Saharan Africa. *African Studies Review,* 42, 15–39.

HelpAge International. (2002). *Leading Global Action on Ageing, Gender and Ageing Briefs.* London: HelpAge International.

Herkovits, M. J. (1937). A note on 'woman marriage' in Dahomey. *Africa, 10*(3), 335–341.

Hobley, C. W. (1910). *Ethnology of A-Kamba and other East African tribes.* Cambridge: Cambridge University Press.

Holling, C.S. (1973). Resilience and stability of ecological systems. *Annual Review of Ecology and Systematics,* 4, 1–23.

Ikdahl, I. (2008). "Go home and clear the conflict": Human rights perspectives on gender and land in Tanzania. In B. Englert and E. Daley (Eds.), *Women's Land Rights and Privatization in Eastern Africa* (pp. 40–60). Woodbridge: James Currey.

Jaetzold, R., Schmidt, H., Hornetz, B., and Shisanya, C. (2006). *Farm Management Handbook of Kenya* (Vol. II C1). Nairobi, Kenya: Ministry of Agriculture.

Kaplan, I. (1984). *Kenya, a country study.* (Farm Management Handbook of Kenya Volume II: Part C East Kenya). Washington, DC: American University, Foreign Area Studies.

Kenya, The Republic of. (2010). *The Constitution of Kenya.* Nairobi, Kenya: National Council for Law Reporting.

Koopman, J.E. (2009). Globalization, gender, and poverty in the Senegal River Valley. *Feminist Economics, 15,* 253–285.

Lekalakala-Mokgele, E. (2011). A literature review of the impact of HIV and AIDS on the role of the elderly in the sub-Saharan African community. *Health SA Gesondheid,* 16, Art#564, 1–6.

Linsk, N.L. and Mason, S. (2004). Stresses on grandparents and other relatives caring for children affected by HIV/AIDS. *Health Social Work,* 29, 127–136.

Lloyd, C. B. and Mensch, B. (1999). *Implications of Formal Schooling for Girls' Transitions to Adulthood in Developing Countries. Critical Perspectives on Schooling and Fertility in the Developing World.* National Research Council. National Academy Press: Washington, D.C.: National Research Council. National Academy Press.

Nyambedha, E.O., Wandibba, S., and Aagaard-Hansen, J. (2003). Changing patterns of orphan care due to the HIV epidemic in Western Kenya. *Social Science and Medicine,* 57, 301–311.

Oguta, T., Omwega, A., and Sehmi, J. (2004, July). *Infant Feeding Alternatives for HIV Positive Mothers in Kenya.* Abstract retrieved from http://www.ennonline.net/fex/22/infant

Omariba, D.W. (2006). Women's educational attainment and intergenerational patterns of fertility behaviour in Kenya. *Journal of Biosocial Science,* 38, 449–479.

Scoones, I. (2009). Livelihoods perspectives and rural development. *Journal of Peasant Studies,* 36, 171–196.

Shumsky, S., Hickey, G.M., Pelletier, B., and Johns, T. (2014). Understanding the contribution of wild edible plants to rural socio-ecological resilience in semi-arid Kenya. *Ecology & Society,* 19, 34.

Tiffen, M., Mortimore, M., and Gichuki, F. (1994). *More People Less Erosion: Environmental Recovery in Kenya.* Nairobi, Kenya: African Centre for Technology Studies.

5 Gendered food- and seed-producing traditions for pearl millet (*Pennisetum glaucum*) and sorghum (*Sorghum bicolor*) in Tharaka-Nithi County, Kenya

Megan Mucioki, Timothy Johns, and Samuel Kimathi Mucioki

Introduction

The origin story of pearl millet, maintained in Tharaka oral history, tells of a man with two wives. One of the wives was not favoured and was chased away with her children into the forest. The forest-banished wife had to find some way to feed her children. One day when she was looking for wild plants to gather near the river she saw a plant with small grains being consumed by birds. The woman decided to gather the grains and feed them to her children. Eventually, her husband ventured into the forest, expecting to find his wife and children expired from starvation. Instead he found them happy and alive, consuming pearl millet. The husband honoured his wife for her resourcefulness and brought her back to his homestead (Respondent #26, personal communication, February 12, 2013).

This tale of the domestication of pearl millet in Tharaka exemplifies the special role women play as skilled farmers and keepers of agrobiodiversity and globally under-represented crops (Abdelali-Martini et al., 2008; Gittinger, Chernick, Horenstein, and Saito, 1990; Gladwin, Thomson, Peterson, and Anderson, 2001; Gurung and Gurung, 2002). The maintenance of seed is central to women's role as food producers (Howard, 2003; World Bank, FAO, and IFAD, 2009) and, furthermore, the freedom (of men and women) to partake in informal seed systems is paramount to the livelihoods of smallholder farming households in Kenya (Muthoni and Nyamongo, 2008; Sperling and McGuire, 2010). Our chapter presents one case of the aforementioned role of women focused in Tharaka, Kenya, where female farmers are keystone keepers and producers of sorghum (*Sorghum bicolor* (L.) Moench) and pearl millet (*Pennisetum glaucum* (L.) R. Br.) grain and seed and related food products.

Finnis (2012) and colleagues analyse what they call marginalised foods or foods that are associated with low status or poverty, rare or expensive, cheap or common, or seasonally limited or perishable (Wilk, 2012). Marginalised foods are often important to specific geographic regions or cultural groups and contribute to dietary diversity and/or local constructs of social systems, rules, or bonds as well as livelihoods (Finnis, 2012). In Kenya, the national production (in tons) of pearl millet and sorghum has been in steady decline since the 1960s (and perhaps longer) (Food and Agriculture Organization of the United Nations Country STAT database, 2014), leading to increasing rarity and decreasing consumption of both crops and under-representation in plant-breeding and seed agribusiness, particularly in the private sector (Dilbone, 2015). This trend is mirrored on a global scale. Khoury et al. (2014) used national per capita food supply data from the Food and Agriculture Organization of the United Nations (FAO) for 152 countries from 1961 to 2009. The results showed that national food supplies have diversified in crop commodities and in the process became more similar and connected in terms of plant genetic resources. Maize, rice, and wheat remained steadily dominant, while soybean, sunflower, palm oil, and rape and mustard made the largest gains; millets, rye, sorghum, yams, cassava, and sweet potatoes had the largest declines.

In contrast, pearl millet and sorghum are still grown by the majority of farmers in Tharaka and are essential components of food and seed systems, farming landscapes, and social systems (Dilbone, 2015). An associated concern related to marginalised foods is the social or economic marginalisation of the people who grow and consume the crops, although there is no clear evidence that links the two (Wilk, 2012). A lack of understanding or support for dietary preferences by agricultural extension, national agendas, and neighbouring regions can in turn move dietary preferences towards mainstream conformity (Wilk, 2012), further aggravate fluctuating chronic and acute challenges, and perpetuate development strategies that ignore mutually dependent traditional crops and gender associations.

In Tharaka sorghum and pearl millet and their related social agroecosystems have been resilient in their ability to persist through environmental stress, change, and uncertainty. When considering the wider framework of this edited collection, this instance of resilience has relevant ties to what Bahadur (2013) presents as effective understanding of social capital, values, and governance structures (i.e., the fine-tuned rules of seed systems and actors involved), community involvement and inclusion of local knowledge (also practices and traditions), preparedness and planning (i.e., systems of seed saving, cropping patterns, and grain storage), and learning (i.e., transfer of knowledge about seed and crops). Complementary to Bahadur's framework is Padmanabhan's (2011) perspective of feminist social-ecological systems which analyses agrobiodiversity as a social-ecological object governed by unique roles of men and women. Padmanabhan considers agrobiodiversity loss an institutional failure and emphasises that it is important to understand how wider institutions interact with local social and ecological systems. For example, the Kenya Seed Act, modelled after the 1991 version of International Union for the Protection of New Varieties of Plants (UPOV), does not mention informal seed systems, the agrobiodiversity that they support, nor the actors involved (Republic of Kenya, 2010). While these are ghost systems[1] in the national legislation, the informal seed system is the dominant method of seed access for most farmers in Kenya (World Bank, 2013). It is clear that in this case farmers have deviated from national rules, developed their own rules, and generated what Padmanabhan calls "deviant informal local institutions" which better meet their own challenges and offer solutions. For Tharaka farmers, "deviant informal local institutions" are essential to seed security and seed sovereignty, in general, and specifically for the continual maintenance of sorghum and pearl millet. Seed sovereignty speaks to farmer control over their seeds through social systems of saving and sharing (Kloppenburg, 2010, 2014), while seed security describes physical access, availability, and utilisation of seed that is of high quality and meets cultural preference (McGuire and Sperling, 2013).

In this chapter a detailed account of processes and actors related to the maintenance of sorghum and pearl millet seed and grain in Tharaka is presented in order to encourage greater emphasis on research on marginalised food crops, acknowledgement of the vital connection between informal seed systems, seed security, and seed sovereignty, and the support and recognition of the agrobiodiversity keepers (in this case primarily women) who are essential to household food security. Thus we argue that Tharaka women's expertise in seed and grain production and food processing of sorghum and pearl millet, coupled with the sharing of expertise and skills with younger women, are vital and lesser-recognised strategies for securing food and seed in Tharaka.

Data collection and analysis

The data presented in this chapter is drawn from a collection of semi-structured interviews and participant observations collected during 14 months of fieldwork (2012–2013) in Tharaka. Forty-three semi-structured interviews were conducted with both male and female respondents (based on Creswell and Miller, 2010). Additionally, four semi-structured interviews were conducted with male and female respondents who no longer maintained pearl millet and sorghum in Machakos and Makueni Counties for comparison (see Eisenhardt, 1989). In some cases, interviews were conducted with the wife and husband together, or paired with other farmers if the respondent had visitors who also wanted to contribute; thus the actual number of people who contributed their knowledge was greater than 43.[2] Interview respondents were selected using purposive and theoretical sampling but not necessarily selected based on gender (Coyne, 1997).

In this chapter we approach our data as a form of narrative because respondents often relayed information through stories and chronological events that moulded their management, access, and continued cultivation of sorghum and pearl millet. According to Elliot (2005) there is not one clear way to analyse narratives, although there are heuristic devices used to engage with the data beyond the surface content. Interview data were transcribed and organised by categorical coding using MAXQDA software, with sorting based on common themes, actions, and experiences in order to make connections among respondents and draw our wider patterns and themes (Elliot, 2005).

The themes of interest for this chapter are linked to observable materials, processes, or behaviours (e.g., seed selection, food preparation, storage, planting). Thus participant observation was used to gather a richer understanding of the lived experience in Tharaka. During this time detailed field notes were taken as well as photographs. Plant walks with knowledgeable respondents, usually those who have spent a considerable time herding,[3] were conducted to collect samples of wild plants mentioned in connection to pearl millet and sorghum during interviews. Plant samples were collected, dried, and pressed (Martin, 2004), and taken to the East Africa Herbarium at the National Museums of Kenya for identification.

One respondent who was particularly knowledgeable was Jane, an elder and active farmer in Tharaka. She is the Athena of agricultural tradition and cultivation in Tharaka, exhibiting strong commitment to agrobiodiversity and traditional crops, exceptional breadth and depth of knowledge, and persistent ties to generational traditions. At the time of this study Jane was in her early 70s and living with her son's family but still farming her own land and maintaining many rare varieties of sorghum and pearl millet. Spending time with Jane added depth and understanding to our data concerning traditional processes and older varieties of crops that other respondents reported they heard about only through elder women or oral history. Jane continues to manage many older varieties of pearl millet and sorghum that other farmers believe are extinct and carries on many traditions which other farmers know of only in their memory. Thus many of the specific examples presented in our data set were given further insight by getting to know Jane and repeatedly visiting her homestead throughout the growing season.

Results

Farmer systems of cultivating pearl millet and sorghum

Jane's pearl millet and sorghum fields are characterised by steady sloping grounds and large rocks wedged between crevices of soil. On this land, Jane successfully cultivates pearl millet and sorghum from top to bottom of the sloping terrain (Figure 5.1). She uses intentional patterns of intraspecific cropping, based on timing of maturity and species intercropping to control erosion. Additionally, she has constructed rock lines perpendicular to her sloping fields to catch organic run off. Planted along these rock lines are pigeon peas (*Cajanus cajan* (L.) Millsp.), while pumpkin (*Cucurbita maxima* (L.) Duchesne) seeds are planted in crevices that gather and hold rainfall and soil between larger rocks. Pearl millet and sorghum are first planted in her fields during the long rains (October–December). The grains are harvested from January–February. On the steepest slope only sorghum is cultivated, as it stays in the field as a ratoon crop through the short dry season and produces again during the short rains (March–May). Pigeon peas also remain in the field, often for a full year, as they offer a large stabilising root system to minimise topsoil erosion from rain and wind.

Management of birds

During the short rains, small grain-eating birds (most commonly the red-billed quelea (*Quelea quelea*) and ploceid weavers (*Ploceus* spp.)) are present in droves. Yield

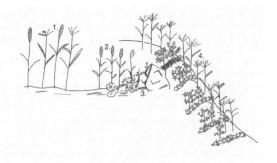

Figure 5.1   Local varieties of sorghum (left) and pearl millet (right) harvested from farmers' fields in Tharaka are displayed in the top two photos. Jane plants her sorghum
and pearl millet on steep rocky slopes, using planting patterns intended to control
erosion and maximise her land. 1. This section is planted only with late-maturing
pearl millet varieties and a sorghum variety, which is grown for market. This
sorghum variety will not be managed as a ratoon crop during the short rains
and can yield even when kept under weeds. 2. This section is planted only with
early maturing pearl millet varieties that require early weeding for good yields.
3. Pumpkins are grown in soil-filled crevices of large rocks. 4. The most sloping area
is planted with a mixture of sorghum varieties alternating with rock terracing lined
with pigeon peas. The pigeon peas and sorghum in this area are managed for two
seasons.

Photos and drawing by Megan Mucioki

loss from birds is considerable enough to deter farmers from growing pearl millet and sorghum during this time. As mentioned above, if sorghum is maintained during the short rains it is usually a ratoon crop from the long rainy season and is maintained for erosion control rather than grain production. Bird scaring is practised during the end of the long rains (December–January) while the densities of bird populations are manageable. The bird population increases with rainfall and food availability and dwindles during the longest dry stretch from May to September. Growing among and adjacent to large rocks on elevated land provides Jane a place for greater perspective of birds encroaching on her field. Bird scaring can be done by anyone in the household but usually is the job of women and children as they are often consistently present around the fields close to the homestead. The labour required to scare birds (see Ainsley and Kosoy, 2015) is the primary reason cited by farmers in Machakos and Makueni for abandonment of pearl millet and sorghum. One respondent, from Machakos, who was about 60 years of age, stopped growing pearl millet ten years ago but still has leftover pearl millet stovers on the roof of his house from his last pearl millet crop, showing signs of its former presence.

Management of off-types

The majority of respondents (women and men) reported two off-types that develop in pearl millet and sorghum throughout the growing season. Women specifically weed out these off-types to ensure quality grain and seed production. In pearl millet the off-type is called *mathara*, which means 'something useless.' Mathara heads are identified by small-sized panicles in length and width and by delicate, shattering seeds; they are weeded out and never selected for seed. In sorghum the off-type is called *munya wa maguna* or 'monkey sorghum.' Farmers refer to this as wild sorghum but it is a weedy intermediate commonly referred to as shatter cane (see Harlan, de Wet, and Price, 1973). Munya wa maguna can grow outside sorghum fields and does not directly develop from sorghum seeds planted by farmers. When munya wa maguna develops in sorghum fields it is directly uprooted because farmers say the plants affect the grain development of their sorghum crop. However, farmers have reported that both these off-types, mathara and munya wa maguna, have been used for food in times of food insecurity. While not a preferred source of sustenance, the off-types have proven resilient in times of extreme environmental stress and, in such cases, useful.

Grain and seed security

Sorghum and pearl millet grain are primarily used for home consumption in Tharaka, although some farmers keep varieties specifically for selling at the market. For example, Jane grows the sorghum variety Mugeta for selling, but keeps the rest of her sorghum varieties (Kaguru, Muruge, Mugana, Muchuri, Kithure, Mweru, Musarama) for home consumption. Additionally, women specifically select a portion of each crop to save as seed for the next growing season. Most pearl millet and

---

Box 5.1 Matrilineal seed system.

Matrilineal seed systems involve informal saving and passing of seed and knowledge through generations of women. In Tharaka, women in each household are responsible for selecting and saving seed each season. Traditionally, when a woman is first married she will receive a portion of seed from her mother and/or mother-in-law. This first seed is a mix of old traditional varieties and includes pearl millet, sorghum, cowpeas, pigeon peas, pumpkins, and melons. After the young woman yields her first crop she returns a portion of sorghum to her mother and a portion of pearl millet to her mother-in-law. Mothers demonstrate the rules of seed selection to their daughters. Unmarried sons often still have their mothers select their seed.

---

sorghum varieties in Tharaka can be found only through farmer seed systems and not in agro shops or through seed aid. Thus, Tharaka women's roles as seed savers of pearl millet and sorghum are paramount to preserving intraspecific diversity as well as the associated traditional knowledge and sentiment (Box 5.1).

For example, one farmer said of his wife,

> *I wanted to drop the Mututwa [local pearl millet] variety from my farm but my wife told me not to drop it because it is the variety that has a good taste for porridge and for everything. So now, last season, because we had not planted a big area, we selected big heads to save for seed and now we want to multiply that this season so that next season we can plant a large portion.*

Grain storage

Jane and her family have maintained traditional grain storage baskets made of wild plants and cow dung, called *mururu* (Figure 5.2; Box 5.2). However, the majority of farmers in Tharaka have abandoned these grain storage vessels for store-bought sacks. Historically, mururu were a symbol of wealth for an onlooker. The large number of mururu signified a household with more crops and land and the ability to trade crops for goats, cattle, or other valuables. According to two elder respondents, households that had 14–30 mururu were considered wealthy.

One male elder explains the cycle of mururu production:

> *What happens if now you have a farm or you are starting to farm, the first thing, you prepare the land for pearl millet and maybe sorghum if you want. And as it is growing, you know that you need to make mururu for you to have a place to store your millet. Then as the millet is growing one could go in the bush and get those materials and start making it as the millet is growing. So when the millet is being harvested he is now finishing the mururu. So when it starts being threshed the millet is then put in that mururu. When the sorghum that has been left there is now maturing he now starts to make another mururu for sorghum. Now the mururu is ready by the time you are threshing*

Figure 5.2 Making a mururu, traditional grain storage container. The bottom of the mururu is started on a flat ground by laying out the munkinyi branches in a circle. Starting from the middle of the branches, nthendu or kagiri are wound from the middle outwards (A). The grasses are secured with mujuria ties (A). This process is similar to starting a basket. After the bottom is complete, it is pegged to the ground and the munkinyi branches are bent upwards into a vertical position and tied into place at the top with mujuria string (B). Grass weaving continues up the sides to the top. Once the grass weaving is finished cow dung is used to plaster the sides. The bottom is plastered first and is done multiple times for better pest control (C). The mururu is filled with grains and for long-term storage (up to 2 years with no pest damage), topped with a cylindrical type calabash, and sealed with cow dung. Those grains which will be consumed immediately are stored in the mururu but the mururu is not sealed shut. Grain is scooped from the mururu using a spoon fashioned from a small calabash attached to a stick.

*sorghum and sorghum is put there. So now as you are increasing your land for farming
you have to make more muruni so that you can be putting your grains in those muruni.
So now after increasing the land that you are farming and you find you are owning more
than 20 muruni.*

Mururu accumulation was a continuous cycle of increasing crop production
and increasing crop storage space. Both women and men could make mururu storage baskets but women were responsible for managing the contents.
Each crop was stored by women separately in a mururu and, depending on the
volume, each variety was stored separately as well. Different sizes of mururu
were made ranging from 5 bags to 50 bags in volume. The real advantage of this
storage method was the capacity for long-term grain storage for times of food
insecurity. This was particularly true with pearl millet and sorghum. Once a
mururu was sealed closed, the grains could remain untouched and unblemished
for 2 years.

Consumption and preparation of grains

The order of consumption of pearl millet and sorghum varieties, once in storage,
is very important for annual cycles of food security. For example, Jane's household
practices the following: First, freshly harvested sorghum requires a period of
"cooling down" or drying before it can be consumed. This sorghum will not taste
or smell appealing and it will not ferment if it is consumed immediately after
harvesting. However, sorghum is also more prone to storage pests. Thus a portion
of pearl millet is consumed first until the sorghum is "cooled down." The rest of
the pearl millet is sealed in a mururu for long-term use. The sorghum varieties
muchuuri and mweru are saved for last among the sorghums because they are least
prone to storage pests. Sorghum is first used to make porridge and some pearl
millet is used simultaneously to make *ugali* (a stiff, dough-like porridge). Once
the sorghum is gone, the pearl millet initially stored in a mururu remains to make
porridge.

Mixing varieties of sorghum in grain storage is dictated by the processing
methods of each variety. Only certain sorghum varieties can be cooked like
rice and others require removal of their seed coat in order to reduce their bitter
taste. The processing methods and associated knowledge are unique to women.
In most instances, men will not consume these small grains if they do not live
in a household where women can process and prepare the pearl millet and
sorghum.

Pearl millet and sorghum are most commonly consumed as different forms of
porridge (Box 5.3). Traditional dishes, such as sorghum pilau or pearl millet ugali
(Box 5.4), in Tharaka, have become rare, having been largely replaced by maize
flour ugali and rice. Additionally, local brew called *marwa* is a popular product of
pearl millet and sorghum. Only women prepare marwa, as it requires preparation of
porridge in the recipe, and porridge itself is made only by women.

Box 5.2 Mururu—Construction and ethnobotanical connections.

Mururu storage vessels are made from a combination of branches and grasses and then covered with cow dung. Mururu are extremely important for long-term grain storage. If sheltered from rain under a constructed shelter, mururu can last and be used for 30 years. Women and men in Tharaka identified multiple small trees, shrubs, and grasses used for making mururu. They are here listed by their local Kitharaka name, scientific name, and use for mururu construction.

Munkinyi (*Abrus schimperi ssp. africana*)—Long, pliable branches are used for the body of the mururu. Wood is valued for termite resistance and bendable qualities. The branches are often lightly burned before use to soften wood and strengthen structure.

Mujirujiru (*Bauhinia tomentosa*)—Used in the same manner as munkinyi in regions of availability. Sometimes it is also called munkinyi.

Mujuria (*Sterculia africana*)—The fresh fibers interior to the outer bark (possibly cambium and vascular tissues) of thin branches are stripped and used like rope for tying down grasses and branches.

Nthendu (*Eragrostis superba*)—A grass used for forming the mururu bottom. It is valued for its ability to bend and merge leaving no holes or cracks.

Kagiri (*Bothriochloa insculpta*)—A grass used to start the bottom of the mururu. It can be used instead of nthendu.

Murunguru (*Hyparrhenia hirta*)—A grass used in the middle and widest section of the mururu. It is valued for its larger length and width. Currently this grass is rare in Tharaka due to grazing cattle and goats. It can be found in some rocky hills that are not exposed to grazing as well as in the highlands outside of Tharaka. This grass is not used today and is replaced with greater volumes of the other grasses.

Kimama Nthia (*Bothriochloa insculpta*)—A grass used following kimama nthia from the neck to the top of the mururu. It is less coarse and sharp, making it better for finishing the top, which will be exposed to people.

Murawa (*Grewia bicolor*)—The thicker ends of branches are cut about four inches long and sharpened at one end. They are then used as pegs and inserted into the ground to hold the bottom of a mururu in place when bending up the munkinyi branches.

Mwathana (*Berchemia discolor*)—Large, straight trunks are used as poles for a simple shelter for a mururu.

Makuri (*Ocimum kilimandscharicum*—Sometimes leaves from this plant are added inside the mururu for additional weevil protection.

Discussion

Food- and seed-producing traditions, linked to sorghum and pearl millet in Tharaka, connect women, plants, and landscapes through processes that have been involved in mediating the in situ preservation of both crops and ultimately the production of seed and food. According to Cannarella and Piccioni (2011) the future of modern

Box 5.3 Mbithi—Traditional porridge.

There are different variations of porridge (mbithi) made in Tharaka, depending on processing techniques, fermentation, and storage vessels. Most people in Tharaka prefer porridge, which has been prepared by hand grinding with the stones (not true in other areas outside of Tharaka). This job is laborious and done only by women. In some cases porridge is made from machine-ground flour. This porridge is usually given to small children during weaning and adults when they are ill, as it is more easily digested. However, machine grinding is not very popular. One respondent from Tharaka explained,

> *People don't like porridge prepared by machines because it makes it to be very smooth. So now machines are not available because those people [mill owners] fear that they will not get customers.*

Traditionally, porridge is flavoured with a wild plant called *rugoiya* (*Indigofera lupatana*) or crushed sweet potatoes. Rugoiya is often preferred only in sorghum porridge but it sometimes makes the porridge thin. Sweet potato is used to sweeten pearl millet porridge. Porridge is stored in a calabash gourd (*Lagenaria siceraria*) smoked with *muthugathuge* (*Solanum tettense*) wood for a valued smoky flavour. It is then left to ferment or combined with fermented milk. Sorghum porridge ferments more slowly than pearl millet porridge. Today, these practices of flavouring and storage have been largely replaced with cane sugar and plastic jerry cans. Those elders who have lived through this transition identify a defining change in porridge taste and quality in correlation with different practices of storage and flavouring.

Box 5.4 Kithongo—A nutritious traditional recipe.

Kithongo is one of several traditional dishes prepared by women in Tharaka that combines locally produced foods into a nutritionally balanced and widely appreciated meal. This meal typically has the consistency of ugali made from maize or other flours. To make kithongo, cowpeas (*Vigna unguiculata*) are added to boiling water; once these are cooked, cowpea leaves, the flesh of *matanka* (citron melon (*Citrullus lanatus var. citroides*), an ancestral form of watermelon), and pearl millet flour are added in the same pot. The dish is stirred until it thickens and is fully cooked. The components of kithongo together make important contributions to daily dietary requirements for energy (as carbohydrate), protein, magnesium, beta-carotene (pro-vitamin A), vitamin C, and B vitamins (including folate, niacin, and thiamin). As the dish is typically eaten, in conjunction with milk (alone or in tea) or with meat stew, meals with kithonga are further balanced in terms of protein and fat content and minerals such as calcium and iron.

The resilience characterized by the diversity of foods and traditional culinary knowledge that ensure nutritional balance is complemented in ecological terms by traditional intercropping patterns. Fields of sorghum, cowpeas, and matanka mirror the well-known complementary corn, beans, and squash of traditional Mesoamerican agriculture and diet (Zizumbo-Villarreal et al., 2012) in their taxonomy (cereal, legume, and cucurbit), spatial architecture, mediation of risk of crop failure, and nutrient output.

sustainable agriculture is linked to "the re-appropriation of lost agricultural traditions or the re-invention of new traditions based on feedback learning processes" (p. 690). Thus, while tradition is rooted in history, it is also dynamic and future-oriented. Specifically in this chapter food- and seed-producing traditions were recorded at various stages of pearl millet and sorghum, grain, and seed production. For example, the ability of Tharaka women and men to identify and collect wild plants, construct traditional grain storage vessels, and follow a special sequence of preservation and consumption builds socio-ecological resilience manifest in prolonged grain stores as an emergency buffer against acute stress; the ability of Tharaka women to master the nuances of traditional recipes that often utilise a diverse range of traditional crops builds resilience through in situ crop preservation and nutritional benefits for women's households. Furthermore, many traditional recipes for sorghum and pearl millet in Tharaka involve lactic acid fermentation, an age-old process that enhances nutrition and taste of the food product and allows for a longer shelf-life (Quave and Pieroni, 2014).

Social capital, values, and governance structures (Bahadur, Ibrahim, and Tanner; 2013) (i.e., the fine-tuned rules of seed systems and actors involved), dictate the organisation and function of informal seed systems and matrilineal seed systems that are vital to accessing seed for both crops in Tharaka. The fundamental dynamics of these seed systems counteract uncertainty and change, as annual seed selection is an iterative, adaptive process based on the interaction of ecological factors and human decisions. Our results recorded the practice of passing on certain types of older varieties of sorghum and pearl millet from elder to younger generations of women during marriage. The pattern is so decisive in Tharaka that we describe this seed system, which is managed by expertise and skills held exclusively by women, and which centres on the passing of these expertise and skills through generations of younger female family members, as a *matrilineal seed system*. Traditionally, this gift of seed set up a relationship of reciprocity in which the bride returned a portion of the seed after the newlyweds cultivated the crops on their own farm. This tradition, once widespread, appears to be limited to only certain households that continue to find the process important to their family; such choices are influenced by the strong presence of female elders, grandparents, or great-grandparents who continue to act as teachers and leaders in their family. Matrilineal systems of reciprocity during and following marriage (as well as abandonment of such systems) have implications for seed security as well as for in situ maintenance and preservation of older landrace varieties of sorghum and pearl millet. Alternatively, cultural practices around seeds can contain an element of 'conservativeness' or adherence to 'old ways' that may not be conducive to the contemporary economic realities or may be seen as an impediment to technological adoption (i.e., new varieties). In this context, on the one hand, change occurring in Tharaka seed systems needs to be explicitly recognised, while on the other hand adoption is more likely if it proceeds within, or does not undermine the social-ecological system that is in place.

Another study in the Tharaka region by Le Clerc and Coppens d'Eeckenbrugge (2012), emphasises patrilocal or virilocal residence—meaning that women in Tharaka leave their clan of birth for that of their husbands upon marriage (Murdock, 1965 [1949]). Presumably, a women would travel with seeds gifted to her, making marriage an opportunity for seeds to travel further geographic distances than characteristic to most informal seed systems (Pautasso et al., 2012). Such a practice (seeds as bride gifts) may have positive implications for the evenness of intraspecific diversity, reducing the risk of some varieties being limited to certain pockets of people or geographic regions, or the risk of extinction from localised disasters. Furthermore, sharing and redistribution of seed resources ensure equity of access to the resources, while community members of different ages and gender have different roles but a common stake in the process of selection and preservation of a resource that most members reply upon. However, Le Clerc and Coppens d'Eeckenbrugge (2012) reported that newly married women received seed only from their mothers-in-law and not their mothers, contradicting our results.

Preparedness and planning (i.e., systems of seed saving, cropping patterns, and grain storage) (Bahadur et al., 2013) are advantageous in informal seed systems, particularly in rural areas like Tharaka, where seed offered through the formal seed sector is often inconsistent, unpredictable, or non-existent (Dilbone, 2015). Special cropping patterns, as demonstrated by Jane's farm fields in our results, require prior knowledge and preparation about the growing potential of specific varieties and techniques for cultivating in semi-arid environments as well as access to the preferred seed in advance to planting. Traditional use of mururu as grain storage containers allows farmers to maintain food stores in preparation for unforeseen stressors as well as for the seasonal lean time. Lastly, selection of off-types designates a set volume of grain for seed and improves the seed quality to facilitate quality crop production in the future.

Learning (i.e., transfer of knowledge about seed and crops) (Bahadur et al., 2013) through matrilineal transfer of knowledge about field management, seed selection, grain storage, and food processing represents a traditional culture, with intergenerational transfer of knowledge ensuring continuation of common values and practices. Local knowledge and subsistence and nutritional values around seed selection and preservation should be respected when new varieties are introduced by agriculture extension and other interventions.

Women's maintenance of seed and knowledge systems is not unique to Tharaka. Eyssartier, Ladio, and Lazado (2008) found that Pilcaniyeu community members in Argentina mentioned (both) parents, followed by mothers and grandmothers, as the dominant transmitters of horticulture-related indigenous technical knowledge to children. However, the researchers reported that horticulture-related indigenous knowledge, more than other knowledge, is in transition because of influences from external groups such as extension agents and researchers. Kerr (2013) cites kin relations as an essential component of food and seed sovereignty in Malawi. For example, relatives and community members possess different levels of obligation in seed giving. Mothers of married women, on the one hand, are obligated to give their daughters seed as a gift if they need seed, and mothers-in-law give daughters-in-law their first seeds to plant after marriage; community members, on the other hand, do not necessarily experience the same societal pressure to share seed. However, the centrality of women to the cultivation and seed-saving of sorghum and pearl millet is unique to Tharaka. A study of gendered labour divisions in a 70 km radius of Douentza Circle in Mali reported that in only one location out of four (Sarafere-Mirion) women were involved in planting millet and men were responsible for all other labour involved with millet (Sperling, 2008). In the other three locations women were not responsible for any aspect of millet production (Sperling, 2008). For sorghum, in these four areas in Mali, only men were involved in the agricultural work (Sperling, 2008).

Lastly, decentralised management of resources and shared cultural values and knowledge are in touch with local realities and enhance community cohesion. At the same time, food security initiatives and economic development increasingly interconnect with county and national institutions. Informal and matrilineal seed systems operate within social, economic, and cultural networks that need to be better recognised in the policy frameworks operating at county and national levels in order to protect these vital sources of knowledge, genetic resources, and food security resilience in semi-arid farming systems. At the same time, policies which may undermine farmer seed sovereignty, particularly for varieties and crops which are not reliably offered through the formal seed sector, require remedial action. More generally, greater effort needs to be made by researchers and policy-makers to better integrate traditional knowledge and practices with technical knowledge from sources outside the community, in order to facilitate household food security in the semi-arid regions of Kenya.

Conclusion

In this chapter we presented qualitative data detailing women's cultivation, management, trading, processing, and serving of pearl millet and sorghum in Tharaka. These two crops are linked to a wealth of information and knowledge related to food producing and demonstrate connectedness to specific types of wild plants that are used in processing porridge and building storage grain containers. Women are the primary controllers of pearl millet grain and seed and possess and transfer related seed knowledge to their daughters and daughters-in-law. Thus we use the term *matrilineal seed systems* to describe seed systems involving crops and knowledge that pass through women's hands.

Resilient components of pearl millet and sorghum food and seed systems in Tharaka include flexibility and diversity; social rules of matrilineal seed systems that support household seed security and transport of intraspecific diversity; social rules that facilitate learning and generational transfer of knowledge; self-sufficiency intrinsic to seed saving and trading that supports preparedness and planning; and diverse cropping landscapes and traditional grain storage baskets, both of which help farmers deal with unpredictable stressors. Utilising data and knowledge about gendered aspects of food and seed systems has the potential to positively impact community involvement in seed variety development and introduction, application of seed policy in rural areas, and decentralised models of formal seed systems. Key local knowledge and agrobiodiversity holders are important contacts for teaching or re-teaching other community members lost knowledge and skills such as mururu-making, seed selection, or cropping techniques. Rare varieties may be successfully multiplied by women offering local agribusiness opportunities if other farmers are interested in reacquiring such varieties. Implementation of such programmes may develop unique opportunities for integrating both indigenous and technical knowledge and practices for promoting resilience and longevity of traditional stewardship of seed and small grain crops by women.

Endnotes

1 We refer to informal seed systems as "ghost systems" in Kenya national seed policy because while policy does not mention or support these seed systems (Tripp, 2006), farmers in Kenya access 80–90% of all cultivated seed through informal seed systems (Almekinders, Louwaars, and Bruijn, 1994).
2 Twenty-seven respondents were female and 22 were male.
3 Herding is often accompanied by the collection of useful plants or wild foods. Thus herders are particularly knowledgeable about local plant identification and classification.

References

Abdelali-Martini, M., Amri, A., Ajlouni, M., Assi, R., Sbieh, Y., and Khnifes, A. (2008). Gender demensions in the conservation and sustainable use of agro-biodiversity in West Asia. *Journal of Socio-Economics, 37*, 365–383. doi:10.1016/j.socec.2007.06.007

Ainsley, M. and Kosoy, N. (2015). The tragedy of bird scaring. *Ecological Economics, 116*, 122–131. doi: 10.1016/j.ecolecon.2015.04.021

Almekinders, C. J. M., Louwaars, N. P., and Bruijn, G. H. De. (1994). Local seed systems and their importance for an improved seed supply in developing countries. *Euphytica, 78*, 207–216. doi: 10.1007/BF00027519

Bahadur, A. V., Ibrahim, M., and Tanner, T. (2013). Characterising resilience: Unpacking the concept for tackling climate change and development. *Climate and Development, 5*(1), 55–65. doi: 10.1080/17565529.2012.762334

Cannarella, C. and Piccioni, V. (2011). Traditiovations: Creating innovation from the past and antique techniques for rural areas. *Technovation, 31*, 689-699. doi: 10.1016/j.technovation.2011.07.005

Coyne, I. T. (1997). Sampling in qualitative research. Purposeful and theoretical sampling: Merging or clear boundaries? *Journal of Advanced Nursing, 26*, 623–630. doi: 10.1046/j.1365-2648.1997.t01-25-00999.x

Creswell, J. W. and Miller, D. L. (2010). Determining validity in qualitative inquiry. *Theory Into Practice, 39*(3), 124–130. doi: 10.1207/s15430421tip3903_2

Dilbone, M. (2015). *Building Seed Sustaining Households: Defining Chronic Seed Security through Informal Seed Systems and Intraspecific Diversity on Semi-Arid, Resource Poor Farms in Kenya* (Unpublished doctoral dissertation). McGill University, Montreal, Quebec, Canada.

Eisenhardt, K. M. (1989). Building theories from case. *Academy of Management Review, 14*, 532–550. Retrieved from http://amr.aom.org/content/current.

Elliot, J. (2005). Interpreting people's stories: Narrative approaches to the analysis of qualitative data. In J. Elliot (Ed.), *Using Narrative in Social Research: Qualitative and Quanitative Approaches* (pp. 50–73). London: Thousand Oaks.

Eyssartier, C., Ladio, A. H., and Lozada, M. (2008). Cultural transmission of traditional knowledge in two populations of north-western Patagonia. *Journal of Ethnobiology and Ethnomedicine, 4*(1), 25. doi: 10.1186/1746-4269-4-25

Food and Agriculture Organization of the United Nations. (2014). CountrySTAT database. Retrieved from http://www.countrystat.org/home.aspx?c=KEN.

Finnis, E. (2012). Introduction. In E. Finnis (Ed.), *Reimagining Marginialized Foods: Global Processes, Local Places* (pp. 1–14). Tuscon, AZ: University of Arizona Press.

Gittinger, J. P., Chernick, S., Horenstein, N. R., and Saito, K. (1990). *Household Food Security and the Role of Women* (World Bank Discussion Papers, 96). Retrieved from: http://www-wds.worldbank.org/external/default/WDSContentServer/IW3P/IB/2000/01/06/000178830_98101903574267/Rendered/PDF/multi_page.pdf

Gladwin, C. H., Thomson, A. M., Peterson, J. S., and Anderson, A. S. (2001). Addressing food security in Africa via multiple livelihood strategies of women farmers. *Food Policy, 26*, 177–207. doi: 10.1016/S0306-9192(00)00045-2

Gurung, B. and Gurung, P. (2002). Addressing food scarcity in marginilized mountain environments: A participatory seed management initiative with women and men in Eastern Nepal. *Mountain Research and Development, 22*, 240–247. doi: 10.1659/0276-4741(2002)022[0240:AFSIMM]2.0.CO;2

Harlan, J. R., de Wet, J. M., and Price, E. G. (1973). Comparative evolution of cereals. *Evolution, 27*, 311–325. Retrieved from http://onlinelibrary.wiley.com/journal/10.1111/(ISSN)1558-5646.

Howard, P. L. (2003). *The Major Importance of 'Minor' Resources: Women and Plant Biodiversity.* (Gatekeepers Series No. 112). International Institute for Environment and Development (IIED). Retrieved from http://pubs.iied.org/pdfs/9282IIED.pdf.

Kerr, R. B. (2013). Seed struggles and food sovereignty in northern Malawi. *Journal of Peasant Studies, 40*, 867–897. doi: 10.1080/03066150.2013.848428

Khoury, C. K., Bjorkman, A. D., Dempewolf, H., Ramirez-Villegas, J., Guarino, L., Jarvis, A. et al. (2014). Increasing homogeneity in global food supplies and the implications for food security. *Proceedings of the National Academy of Sciences, 111*(11), 4001–4006.

Kloppenburg, J. (2014). Re-purposing the master's tools: The open-source seed initiative and the struggle for seed sovereignty. *Journal of Peasant Studies, 41*, 1225–1246. doi: 10.1080/03066150.2013.875897

Kloppenburg, J. (2010). Impeding dispossession, enabling repossession: Biological open source and the recovery of seed sovereignty. *Journal of Agrarian Change, 10*, 367–388. doi: 10.1111/j.1471-0366.2010.00275.x

Leclerc, C. and Coppens d'Eeckenbrugge, G. (2012). Social organization of crop genetic diversity: The G x E x S model. *Diversity, 4*(4), 1–32. doi: 10.3390/d4010001

Martin, G. (2004). *Ethnobotany: A Methods Manual.* London: Earthscan.

McGuire, S. and Sperling, L. (2013). Making seed systems more resilient to stress. *Global Environmental Change, 23*, 644–653. doi: 10.1016/j.gloenvcha.2013.02.001

Murdock, G.P. (1965 [1949]). *Social Structure.* New York, NY: Macmillan.

Muthoni, J. and Nyamongo, D. O. (2008). Seed systems in Kenya and their relationship to on-farm conservation of food crops. *Journal of New Seeds, 9*, 330–342. doi: 10.1080/15228860802492273

Padmanabhan, M. (2011). Women and men as conservers, users and managers of agrobio-diversity: A feminist social-ecological approach. *Journal of Socio-Economics, 40*, 968–976. doi: 10.1016/j.socec.2011.08.021

Pautasso, M., Aistara, G., Barnaud, A., Caillon, S., Clouvel, P., Coomes, O.T. et al. (2013). Seed exchange networks for agrobiodiversity conservation: A review. *Agronomy for Sustainable Development, 33*(1), 151–175. doi: 10.1007/s13593-012-0089-6

Quave, C. L. and Pieroni, A. (2014). Fermented foods for food security and food sovereignty in the Balkans: A case study of the Gorani people of northeastern Albania. *Journal of Ethnobiology, 34*(1), 28–43. doi:10.2993/0278-0771-34.1.28

Republic of Kenya. (2010). *National seed policy*. Nairobi, Kenya.

Sperling, L. (2008). *When Disaster Strikes: A Guide to Accessing Seed System Security*. Cali, Columbia: International Center for Tropical Agriculture. Retrieved from http://pdf. usaid.gov/pdf_docs/pbaaa614.pdf

Sperling, L. and McGuire, M. (2010). Understanding and strengthening informal seed markets. *Exploratory Agriculture, 46*, 119–136. doi:10.1017/S0014479709991074

Tripp, R. (2006). Strategies for seed system development in Sub-Sahara Africa: A study of Kenya, Malawi, Zambia, and Zimbabwe. *ICRISAT, 2*(1), 5-50. Retrieved from: http:// ejournal.icrisat.org

Wilk, R. (2012). Loving people, hating what they eat. In E. Finnis (Ed.), *Reimagining marginialized foods: Global processes, local places* (pp. 15–33). Tuscon, AZ: University of Arizona Press.

World Bank. (2013). *Agribusiness Indicators: Kenya*. Washington, DC: World Bank.

World Bank, FAO, and IFAD. (2009). *Gender in Agriculture Sourcebook*. Washington, DC: World Bank. Retrieved from http://elibrary.worldbank.org/doi/ pdf/10.1596/978-0-8213-7587-7.

Zizumbo-Villarreal, D., Flores-Silva, A., and Colunga-Garcia Marin, P. (2012). The Archaic diet in Mesoamerica: Incentive for milpa development and species domestication. *Economic Botany, 66*, 328–343. doi: 10.1007/s12231-012-9212-5

6 Banking on change

An ethnographic exploration into rural finance as a gendered resilience practice among smallholders

Carly James and June Y. T. Po

The young, soft-spoken manager of the bank sat calmly in his front office. It was the middle of the dry season in Makueni County, leaving temperatures fairly mild during the day but very cool at night. "You see, banking is really all about trust," he explained. "It's also a matter of discipline." Here was the celebrated new manager for the local branch of K-Rep (Kenyan Rural Enterprise Program), Kenya's oldest commercial microlending institution, which specifically targets the rural and urban poor. He was in his late 20s and maintained a high-energy, eager tone as we spoke. We sat in his office talking for several hours as clients filed through the modest front lobby to the tellers with their passbooks—palm-sized notebooks given to all clients for accounting—in hand, some being referred directly to the manager himself if they had specific questions or issues. After a while, it became evident that the clientele who came into the office that day was exclusively female. After being prompted, the manager explained how one person would often come into the office as the representative of a lending group comprised of five to ten members (mostly women). While K-Rep began as an individualistic microlending institution, the manager explained how K-Rep's community development arm decided to expand its practice to incorporate a new group-lending model after seeing how well it worked in other countries.[1]

This particular K-Rep branch is located in the bustling town of Wote, capital city of Makueni County, located about 140 km southeast of Nairobi, in the semi-arid midlands of Kenya's Eastern region. The Ákámbá (Kamba) and other local peoples lead a predominantly agrarian lifestyle in spite of their challenging, low-rainfall environment. Kamba smallholder farmers generally rely on subsistence farming, combining food crop and livestock production under conditions of moderate land use intensity (Jaetzold, Schmidt, Hornetz, and Shisanya, 2006). Besides subsistence farming, many farmers engage in casual labour and non-farm activities. Women work on the farm or engage in artisanal income-generating activities such as basket and rope weaving, or managing a fruit and vegetable kiosk or tea shop in the local market. With low annual rainfall, long dry seasons, and poor rural infrastructure, agricultural finance is considered a high-risk venture for both debtors and lenders. Still, in times of economic and climatic instability, it appears that the financial practices of smallholders remain relatively steady from season to season. Many farmers are involved in informal, interpersonal finance—whether through peer-to-peer borrowing/lending, group-based money-sharing, or otherwise—while only a minority transact financial matters through formal, registered financial institutions. These levels of activity do not appear to fluctuate according to bad harvests or drought. Instead, smallholders in this area are motivated to explore new financial outlets (or deepen their existing financial commitments) due to a variety of reasons that often do not directly relate to the harsh environmental challenges they face from year to year. As one farmer explained, "We do not get up and leave just because it is a bad season. These [credit/savings] groups are the way we stick together."

> *"Giving a farmer a loan of KSh 50,000 (USD $570) is like giving a child githeri (traditional maize and beans dish): they can't chew it."*
> —*Local bank manager, Wote, Eastern Province*

The global microcredit industry has mushroomed in the past several decades. In some sense, this trend represents a wider shift in development rationality from the state-led provision of financial resources for the underserved to the expectation that individuals who are aware of their lesser financial standing will act responsibly to secure their own well-being (Rankin, 2001). 'Microcredit' has become a buzzword among lending institutions in Ukambani. What is less certain is whether or not microcredit, as a movement in global development and as a particular mode of formal lending, has the same distinction and uniqueness of appeal to farmers on the ground. What is clear, however, is that farmers draw harsh distinctions between formal and informal forms of credit—and they do so strategically.

Introduction

In this chapter, we review rural sources of credit and credit's relative importance in the smallholder farming system. We investigate how the landscape of credit sources may be gendered, and how access to such credit sources may differ according to the gender of the borrowers. Because this chapter posits credit as an important resource for enhancing the socio-ecological resilience of the smallholder farming system, it also examines the inverse: that is, to what extent does the smallholder farming system, with its range of collective activities and group-based livelihood strategies, contribute to the establishment of credit-worthiness in both formal banking institutions and informal savings groups, and, moreover, how group-based banking may make credit fairly and readily accessible to the majority of farmers. A gendered perspective is key to this analysis, since it has been demonstrated that giving women the same access to productive resources, technologies, and services as men tends to increase agricultural productivity and, ultimately, household food security and general welfare (Quisumbing and McClafferty, 2006).

The chapter relies primarily on 35 household interviews, several focus-group discussions, and interactive exercises (June to August 2013) organised by the authors around the topic of credit and financial strategies of women and men. In addition, the chapter draws on insights from a further 75 in-depth interviews, 16 focus groups, and 7 community meetings (2013 and 2014) focused on smallholder farmers' access to land resources in semi-arid Makueni County. Our analysis employs an ethnographic perspective on credit, but also is informed by discussions of credit in other disciplines (e.g., credit as part of the domestic economy). The ethnographic perspective allows for a rich and nuanced understanding of the ways smallholder farmers conceptually engage with the world.

The overarching goal of this chapter is to demonstrate, through ethnographic evidence, how formal and informal credit is a valuable component of the smallholder farming system, especially insofar as it has the capacity to contribute to a household's food security resilience. The important takeaways from the body of research presented here pertain to the ways in which farmers are confronting barriers to accessing credit, how these barriers sometimes differ in relation to the gender of the borrower (based on financial literacy, financial management practices, and more), and the common ways forward in informal lending practices. Credit is a livelihood resource. Our analysis in this chapter examines issues of gendered access and entitlement to this resource, whether from sources formal or informal. Navigating the gendered landscape in which credit operates paves a way forward for understanding how to improve upon financial markets in order to make them more relevant for women and for a resilient farming system.

Imperfect money markets

> *"Microcredit has been identified as the governmental technology most suited to the objective of building rural financial markets."*
>
> —*Katherine Rankin, Governing Development*

Ukambani is a semi-arid region where soil erosion and climatic variability are making financing any level of commercial farming a precarious venture. Moreover, developing sources of credit that are consonant with smallholders' seasonally dependent needs often translates to low profitability for lending institutions. In the same way, the farmer's perspective tells us that the transaction costs and added fees associated with travelling to a lending institution, submitting a loan application, and waiting for it to be processed are simply not congruent with the enormous time and money constraints that bear on most farmers. Chronic food insecurity is an ongoing feature of the smallholder livelihood system in Ukambani which layers urgency on top of already pressing sets of credit needs. Smallholder credit needs are unique in that they generally require smaller amounts of money (by most standards, 'microloans') more frequently, albeit not always regularly. Lending institutions often struggle to meet these needs in a cost-effective way. With this in mind, this chapter addresses the question of where farmers turn for credit in a time of need. And to what extent does credit offer the smallholder some predictability amidst a very unpredictable and challenging environment?

In Ukambani, 'microcredit' can mean a number of things. 'Microfinance institutions' refer to the formal providers of small-scale financial products. Not all microfinance institutions offer the same services. Some may offer clients checking accounts, some may lend only to groups instead of individuals and some may lend in-kind loans rather than cash. In terms of sheer physical presence in the most high-traffic areas, the four microfinance institutions which dominate the landscape in our field site of Wote Town in Makueni County include K-Rep, Kenya Women's Finance Trust (KWFT), Small and Medium Enterprise Programme Microfinance Bank (SMEP), and Faulu. 'Microcredit' as a concept can refer to the centuries-old practice of peer-to-peer indebtedness, whether through group lending practices or individual-to-individual. This kind of small-scale lending, couched in social sanctions, strengthens farmers' ability to respond to environmental shocks in Ukambani. As Robinson (2001) notes, this relatively simple form of informal microfinance does not appear in state-level calculations of Kenyans' financial activity, making it appear as if the poor go 'without' credit most of the time. A closer look, however, reveals extensive webs of indebtedness among the rural poor where local understandings of morality, materiality, and processes of development play out (Robinson, 2001, p. xxxi).

Money matters: The role of credit in smallholder farming systems

"Money is always there but the pockets change; it is not in the same pockets after a change, and that is all there is to say about money."
—*Gertrude Stein*, Saturday Evening Post, *June 1936*

In the wake of what is often framed as a progressive 2010 constitution, Kenya's Vision 2030[2] development plan is making waves with major structural changes that are meant to promote vast economic growth for the country. Among these changes are heightened levels of financial support for smallholder agriculture in the country's semi-arid regions, where populations tend to be more 'unbanked'[3] than elsewhere in Kenya. In particular, the Central Bank of Kenya is promoting the "financial inclusion" of more poor and low-income Kenyans into the formal financial sector (Anyanzwa, 2012). The formal financial sector refers to the taxed, regulated segment of the financial activity of all Kenyans. Because formal transactions tend to occur through existing channels in commercial lending institutions or the government, formal finance also has the connotation of being traceable. Informal finance, however, is usually framed as being less transparent. Important to note here is the fact that smallholders are statistically more active in informal (non-regulated, non-taxed, not incorporated into official calculations of gross domestic product) finance. Informal finance has been characterised by some as a method for circumventing or even undermining the institutional pathways of state-regulated finance, but the reader will see in this chapter that most commonly, informal finance is appealing to smallholders for more pragmatic reasons. That is, rather than actively resisting formal finance, farmers who seek informal credit do so typically out of convenience, accessibility, and historical precedence.

Available statistics from the Central Bank show that 35.2% of Kenyans are unable to access formal financial services.[4] As Meier and Rauch (2000, p. 317) argue, however, "low-income consumers do not simply consume less: they consume goods and services that serve similar purposes but at a much lower price." In Kenya, the informal sector (including informal sources of finance, informal sources of employment, etc.) is quite large, estimated at 34.3% and accounting for 77% of employment. Over 60% of those working in the informal sector are youth (ages 18–35), split fairly evenly between genders (Ouma, Njeru, Kamau, Khainga, and Kiriga, 2007). In Ukambani, certain traditional artisanal activities have become important sources of income, although they are practised informally through cash-only transactions and brokering that is not necessarily recognised by the government—such activities may include ebony carving, gravel making (that typically occurs roadside, to reduce transaction costs for consumers), and basket weaving using sisal fibre.[5] Many of the smallholder farmers in this area also participate in "casual" labour (e.g., housework, carpentry) in the off-season that is rarely documented as a legitimate source of income. While women who are engaged in paid work may typically run small stalls (e.g., tea shops, hair salons) in town or do artisanal work such as rope-making, men also commonly seek employment in a variety of ways: as casual labourers; collecting and selling water to neighbouring farms, or local schools; delivering goods to markets (if they own a donkey or ox-drawn cart); operating a motorbike business, or renting farms to cultivate cash crops such as tomatoes and kale (personal observations, 2013).

Group lending has become a dominant theme in the informal financial sector, especially in the rural areas where it can be difficult to access banking institutions. In Ukambani, it is very rare to meet a farmer who is not involved in at least one, if not multiple, self-help groups.[6] Today, with the growth of the cash economy and higher rates of mobility, we see people forming groups for the purpose of exchanging not only money but also materials, time, and even knowledge and skills. Since precolonial times, the Kamba have participated in *mwethiya* (self-organised collective labour groups, also variably identified as using the Kiswahili *chama*) where farmers pool labour resources and collectively till the land, plant, herd, and the like. Without delving into the historical background on how these groups arose and why they became so entrenched in everyday life, here it is worth describing the way in which these groups operate and the position they hold in farmers' financial lives. Typically occurring in very regular intervals (e.g., weekly, monthly), group members make regular monetary contributions to the group 'pot' which is then loaned to one member for a specified amount of time; at the specified time, the group member returns the loan in full, with interest, at which time the loan is given out to another member, and so on. This rotating structure is often credited as the model's core strength, since it ensures that each group member is assisted in due time—as a result of this mutual assurance, many people join mwethiya even if they do not have immediate, specific, or known needs (Ardener and Burman, 1996; Ochieng and Maxon, 1992).

Framing credit as a constraint on smallholder livelihoods, as the Kenyan state has done through financial inclusion policies, belies the productivity of creditor-debtor relations that occur informally (Freeman, Ehui, and Jabbar, 1998). In a March 2014 article from one of Kenya's largest newspapers, the *Standard*, more than 1.4 million Kenyans were reported as relying exclusively on "shylocks and friends" for loans, which was deemed wholesale as "unsettling."[7] Transactions made outside of the framework of formal calculations and documentation—informal banking—is dismissed as a subversive market, a lesser option, or a last resort alternative to formal banking by the government and dominant development narratives promoting modern statecraft (Scott, 1998). The lived experiences of Kamba farmers provide an alternative perspective which reveals the fact that informal credit often plays an important role in enhancing the socio-ecological resilience of the smallholder farming system. Informal credit allows farmers to have cheap, direct, reliable access to credit through their social networks. Our ethnographic research supports the idea that interpersonal transactions enhance solidarity within a community, which leads to greater resilience. Webs of transactions, indebtedness, and lending between and among people statistically have virtually non-existent default rates. They bring people closer together, and they ensure that there is an inter-reliability among members of the community. This is not to make any sweeping moralising statements about what is the best or right option for small-holder farmers, but rather, the ethnographic evidence in this study does expose the efficiencies of the informal sector for strengthening smallholder livelihoods, in particular by bolstering community inclusion and preparedness for building a more resilient farming system.

Barriers and constraints to accessing microcredit

"Farmers who can get credit generally don't need it."

—*John Gerhart, 1975*

In the introduction to this volume, the editors highlight the idea that *vulnerability* may be thought of as *resilience*'s antonym. A system with vulnerabilities has a lower capacity "to adapt to and shape change," which, in smallholder farming systems, can have devastating consequences in terms of hunger, malnutrition, and general well-being (Folke et al., 2002, p. 13). In this portion of the chapter, we follow this line of reasoning to posit that when access to appropriate sources of credit is impeded or altogether absent or blocked, this may be conceptualised as creating vulnerability for farmers within the smallholder farming system. At the same time, we want to drive home the point that credit, when an appropriate source thereof is actually secured, has the capacity to enhance the resilience of the smallholder farm-ing system. Below, we review some of the barriers and obstacles that farmers face when accessing credit.

One thing that was often cited as an obstacle during household interviews with farmers was collateral requirements. In her research, Susan Johnson (2004) determined that local land and inheritance rights were one of the most important factors in explaining why using title deeds as collateral is a barrier for farmers who seek credit from a formal financial institution. She explains that "[t]he social relations surrounding [land] produce a situation where its use as collateral is heavily constrained," which explains why she found that the overall use of bank loans was relatively low in her Kikuyu field site in Central Kenya. Moreover, Johnson explains that Kikuyu husbands are encouraged to consult wives and other family members in the process of mortgaging land, and she notes that these family members were unlikely to agree. Those microfinance institutions in Kenya that rely heavily on a group lending structure (e.g., Faulu, KWFT, K-Rep FSA, SMEP) make use of the "social technology of joint accountability" where the group members actually guarantee the others' character (Johnson, 2013, p. 64). This usually requires a certain amount of paperwork and household visits to demonstrate that the person will serve as a reliable borrower. However, it remains much more cost-effective as an accountability mechanism than putting forth major assets as collateral. These institutions are taking a tip from informal microlending, where social sanctions have long been used as an incentivizing tool. Typically, groups or individuals rely on social connectedness as an accountability measure against defaulting—i.e., 'saving face' among friends, or what some have termed 'reputational collateral.' Using collateral that is rooted in social norms also provides an opportunity for farmers without a credit history to become borrowers and lenders. Several farmers in our study indicated that they were uninterested in taking loans with banks or microfinance institutions because they knew that their credit history could be looked up, and it made them feel vulnerable. When they lend with their friends and family, they said, it allows them to do things on their own terms. As one farmer said, "They know us. They know our problems."

Within the Kamba family unit, both men and women are expected to work the farm, where women are entitled to cultivate and benefit from the farm food crop and sales of harvest. Men are entitled within Kamba traditions to inherit a portion of their father's farms as owners. As traditional inheritance of land is shifting towards formal ownership of land as freeholdings, this has greatly impacted the ability for smallholder farmers to acquire loans. With ongoing national and local pressures toward land adjudication and titling, some farmers in Watema, Mumbuni and Kathonzweni (all in Makueni County) have been issued title deeds. If the family possesses a title deed for the land, the cultural understanding is that the wife can transfer her name onto the title deed after her husband dies. Until then, however, the male head of household is the proprietor of the land. Similar to what Johnson (2004) found in Central Kenya, social norms surrounding land rights sometimes keep people from securing a loan by offering their title deed as collateral.

There is sometimes tension between grandfathers and fathers when the grandfathers are the ones who acquired the title deeds but are very reluctant to subdivide the land formally to their sons. "There are those people who don't want

[to] give out the title deed; the grandfather is the one who [subdivides]." Among the Kamba, the sons tend to want access to the land so that they can productively use it. From their father's perspective, the land is typically best looked after by the father until the son is married and has a pressing need to make farming a key part of his livelihood. Several farmers indicated that the older Kamba men are worried that all the young Kamba men are running off to the urban centres to find better work, which makes the older Kambas more eager to control the land themselves. Apart from cultural and social norms, the administrative fees for issuing or transferring the title deed are prohibitively high for many farmers. Informal verbal instructions on land subdivision are the norm, yet when left undone before the title holder's death, leave family members with potential conflicts and family disputes.

In an interview with the Wote branch manager of the Agricultural Finance Corporation (AFC), an institution wholly owned by the government, we learned that about 80% of AFC's clientele are men due to the fact that the only form of collateral AFC accepts is a title deed.[8] Importantly, he was quick to note that AFC is simply "not a bank" and therefore reserves the right to exercise certain restrictions on who they target and accept as clients.[9] When asked about the difference between banks and non-bank microfinance institutions, Makueni farmers were pretty clear that they understood them all as banks. That is, they indicated that there is no linguistic distinction between a "commercial bank," a "village bank," or a "microfinance institution" in Kikamba, since all of these terms are categorically described using the same term: *vengi* (banks). One farmer even said outright, "They are all banks to us," indicating that the very term *venga* (bank) may reflect a social norm of avoiding the formal sector altogether. Moreover, this reinforces how the farmers think of the formal sector as a monolithic whole, indicating that one barrier to accessing formal microcredit could simply be psychological. Historically, Kamba farmers have not turned to banks for their credit needs.

Whether there are microcredit institutions trying to market directly to them or not, it is clear that Kamba farmers do not consider "banks" to be part of their everyday lives. But perhaps there is more to the idea that farmers in Ukambani are not turning to banks out of habit or custom, meaning there is a deliberate avoidance that is made more convenient with universalising terms like *venga*. Many informants indicated that formal institutions are generally quite intimidating and confusing. In the words of one farmer, "You feel like you're being duped." There were quite a few bureaucratic elements that seemed to discourage people from borrowing, such as legal jargon and added costs for "processing" and having one's application stamped by a lawyer. Farmers cited the large loan sizes and high interest rates of up to 36% as other reasons that discouraged them from borrowing from these institutions. The gender bias of formal lending institutions is readily apparent, with much of the documentation requiring advanced-level reading skills. Kamba women are less likely than men to be literate and the bureaucratized nature of formal banking emphasises these disparities. Women cited various other reasons why they did not like to approach banks for credit, including the time it takes to process a loan application, the collateral requirements (men are usually the formal owners of the kind of high-value assets that are required as guarantees on taking a loan), the impersonal feeling of banks, and the reading of dense texts that seem to raise suspicions. Most commercial banks (excluding microfinance institutions that specifically target women, such as Faula or K-Rep) require a husband's signature even if the loan is requested by a woman. One woman explained, "[A]t the end of the day, women in Africa don't own anything."

Due diligence: Women smallholders' experiences with credit

> *"Relatively speaking, hunger and poverty are more women's issues than male issues. Women experience hunger and poverty in much more intense ways than men."*
> —*Muhammad Yunus*, Banker to the Poor *(1998)*

A growing number of Kenya's microfinance institutions target women directly. From the formal financial sector's perspective in Ukambani, microcredit connotes a clientele comprised mostly of rural women. From the perspective of local farmers, women are clearly much more involved in informal group lending than in any other mechanism of accessing credit and saving. Based on our ethnographic research alone, it is clear that women tend to be more 'unbanked' than men by far, historically and conventionally tying themselves to informal, localized group lending practices.[10] Women also tend to interact with and control less income than men overall, with most Kamba women that we spoke to indicating that they manage expenses within the household (primarily food and anything to do with the children), while men focus on the larger purchases that are directly linked to their livelihoods. In part, this contributes to the dominance of women in informal group lending where microloans are the norm and most actors occupy the same low-income level. Still, some more affluent women do turn to microfinance institutions as a way of accessing money when in need. Most often, their interactions with these institutions came in the form of shareholder group settings.[11] As it was explained to us in our interviews with farmers, microlending groups were a way of "sticking together" and relying on his/her peers in times of need. Debt relations (i.e., where someone is a debtor to someone else) are intensely personal in Ukambani, so people are acutely aware of the problems that their peers are facing and know how to react accordingly. In Ngtini Village, nestled in the hills of Kivani, we spoke with farmers who were involved in groups that were cleverly titled to reflect their aims of mutual assistance, such as Mbitike Ngwitike ("Call me, I'll respond") and Kanini Kaseo ("Small, but good"). In Mumbuni area, microlending groups took names such as Koma Wisi ("Sleep satisfied") and Umoja Group ("Unity group"). The reasons women joined these groups varied, but centred on a theme of helping one another. One informant claimed, "When you're in a group, you guarantee each other."

Farmers cited benefits of self-help group involvement that pertained to activities outside of the group setting. The members of these groups benefit from having closer connections and identifying shared points of interest outside the scope of the group. For instance, many of the women we spoke with were involved in political campaigning for the same candidates, saying that the reason they first met and talked about their shared political interests was because of their memberships in various groups. There were also cases of conflict resolution taking place within the group setting. One farmer in Ngitini Village talked about the way his involvement in self-help groups had helped him gain some prominence as a community leader: "I do bring people together so that we are one. Yes. Have you seen? If something has come up, we sit and talk about it. And if someone is missing, I will look for him or her and tell him or her about it so that he or she can evaluate and accept or refuse it." He indicated that these groups promoted in him a sense of responsibility and duty to his peers. Box 6.1 offers an ethnographic snapshot of one of the many ways that these deepened social relations play out.

In Bangladesh, research has shown that microfinance can have an impact on gender relations within the household, as women are able to better support the family as a whole through their earnings (Schuler, Hashemi, Riley, and Akhter, 1996). At the same time, conflicting reports of women's financial "empowerment" through microloans have shown that in some instances, microfinance increases a woman's bargaining power within the household to the extent that a private engagement with money has resulted in cases of increased domestic violence against women out of accusations of secrecy or jealousy (Johnson, 2004). Anthropologist Parker Shipton (2010) highlights the "supply-led pattern" that plagues much of

---

Box 6.1 Charity's case

Charity, a mother in her 60s, borrowed a two-year loan from K-Rep microfinance to start a fish-farming business. She dug the pond in her farm and bought a plastic lining for the pond. Unfortunately, the birds in the area punctured the lining during the stages of preparation and water could no longer be retained in the pond. Unable to repay her loan within the allotted time, she sought help from her K-Rep women's savings group. The loan came with an agreement from her group that their savings account containing 300,000 Ksh would be frozen until her loan was repaid. K-Rep had agreed to extend the repayment period. The group of 25 women members contributed 200 shillings each per month for 5 years to the savings account. Since then, members of the group had come to Charity to ask for repayment of the loan in order for them to reopen their savings account, but she had two children in secondary school, for whom she diverted household income to pay school fees. Until her children complete secondary school, she would not have the income to repay the loan. The last time the group members came to ask for repayment was in 2012. The group no longer contributed money to the savings account since it was frozen. Although she still met her K-Rep group members within the community, they had left the repayment issue as it stood. Although the group members were disappointed by her inability to repay, they remain connected in other ways and they understood Charity's situation. "I was saved by the group," Charity said. She hoped to rebuild her pond in the future. She heard that World Vision in the area was providing training and free supplies for farmers to start their fish-farming business. Even though she was not a part of the World Vision group, she remained hopeful.

the lending scene in Kenya, where for political reasons or otherwise, support for rural lending is done through subsidizing institutions that have not necessarily proven to be popular by demand. For women smallholders, this may mean that support is given to institutions they may not even be able to access (e.g., due to poor infrastructure or confusing legal requirements), let alone use regularly out of habit or convenience. Due to their ostensible command over the farm,[12] men often seek loans seasonally and in larger volumes than the loans women typically handle. Commercial lending institutions tend to favour debtors who require larger volume loans, as bankers reported that this connotes a weightier responsibility on the debtor to return the money on time.

In Ukambani, it is clear that women control less income than men, but some women are consulted in decision-making processes about consumption relating to the farm. When it came to decisions about household consumption, women typically said that what money they were in charge of they spent on the household (e.g., school uniforms, pot and pans, or in Kiswahili, *sufuriyas*[13]) and that they made such decisions unilaterally. Still, women reported not having direct control over money—if women did receive small amounts of money from peers or groups, some preferred to handle their money in secrecy, away from their husbands, to secure it against theft. Men in Makueni County generally reported that the money they received from petty trading and the selling of agricultural products first came to them, and then they would give it to their wives in the form of petty cash or allowances for grocery items (e.g., sugar for tea, flour, diapers). Sometimes, wives reported having to ask their husbands for money for basic items and other times; wives were given relatively small allowances of cash regularly.

Part of the Ukambani norms around marriage gives wives the entitlement to ask for money from their husbands. "Husbands are the ones responsible for bringing food home" in the form of income, parallel to the traditionally Western concept of "the breadwinner." Often women exercise this entitlement selectively. If they ask for money for household expenditures too frequently, they would be accused of squandering the money on frivolous items like extra clothing. "The husband will ask me, 'Where is the money I gave you yesterday?'"

Conclusion: Resilience and smallholder credit

> *"Intimacy colors entrustment and obligation. People who borrow and lend, as we all do, tailor the terms of their loans and repayments according to interpersonal relationships."*
> —*Parker Shipton*, Credit between Cultures *(2010)*

If credit plays a productive role in making the smallholder farming system more socially and ecologically resilient, then the system can only get increasingly more stable when sources of credit are couched in social terms. Looking at the way credit and debt tie people together is a way of understanding the smallholder farming system's resilience characteristics, since debt relations do strengthen social relations by deepening and broadening people's existing social networks and ensuring that everyone has someone else to turn to in a time of need. In particular, lending through group settings—whether through shareholder groups linked to commercial banking institutions or through more informal self-help groups—have been studied extensively as being more effective in strengthening community bonds and building solidarity than being indebted individually to an institution (Ardener, 1964; Ardener and Burman, 1996; Bouman, 1995). This is especially true in more communal cultures and in rural areas, such in Ukambani (Kinyanjui, 2012).

In summary, farmers in Ukambani appear to be most interested in informal finance, where debt is understood as a positive economic indicator because it brings people together. Perhaps investments made into supporting smallholder agriculture should more intensely focus on the merits of informal modes of lending. More than that, looking at these issues from the woman's perspective on credit allows us to also see the ways in which the genders favour different sources of credit and whether or not this diversity of sources of finance is effective in enhancing the resilience of the smallholder farming system.

When it comes to analysing the effects of credit on the smallholder farming system, we invoke the work of Akin and Robbins (1999) in which researchers were encouraged to make distinctions between actors' "transactions for personal gain and those aimed at social reproduction," since not all credit relations are created equal. Certainly along gendered lines, interactions with credit differ because women and men farmers often have dissimilar seasonal credit needs (with men typically having large-scale credit needs during the planting season and women needing smaller-scale loans to pay for school uniforms early in the year and occasional food items like special flour for Christmas chapati). Consistent with Susan Johnson's work in Central Kenya (2004), our research has found that women tend to operate in smaller-sized amounts of money and have more frequent credit needs since they are focused more on the everyday needs of the household. Men, on the other hand, tend to desire larger-sized loans and live their financial lives more seasonally according to the needs of the farm. Bearing this in mind, many microfinance institutions market their products and services directly to women, following Vision 2030's demands for financial inclusion; but this elides the number of obstacles that end up preventing women farmers from actually turning to formal microfinance institutions in a time of need. Among these are processing and legal fees, a general fear of dispossession and public shaming, and prohibitively high interest rates on individualistic/private loans that do not offer the social benefits that informal groups might. While the need for financial assistance is evident, farmers consistently expressed suspicion with regard to taking loans from formal financial institutions (e.g., commercial banks, microfinance institutions). Many farmers expressed disaffection with banking institutions and the concern that defaulting on a loan would result in material dispossession.

Our research also finds that farmers use different sources of credit in different ways. They do not seem to perceive income-generating or commercial activities to be the best or only pathway for making their livelihood system more resilient against shocks. While bank loans are often marketed as a method for supporting income-generating activities alone, credit from informal credit/savings groups is often used to pay school fees, purchase foodstuffs, or for other ad hoc expenses—items that do not typically qualify as 'investments' from which one might expect an immediate, productive return (Karim, 2011). Women's focus on supporting the domestic economy is one of many ways of bolstering the integrity of the entire smallholder farming system, because it supports the immediate and long-term health and well-being of the family.

Because smallholder livelihoods are both subsistence-oriented and commercial in nature, the maintenance of the household comprises a set of needs which are arguably just as pressing as those of the farm. If a man receives farming advice from his peers as part of a self-help group, he is better equipped to tolerate shocks to his livelihood. Equally, if a woman receives a loan from her mwethiya and purchases school uniforms for her children, she also contributes to the resilience of her livelihood by ensuring that the household needs are fulfilled. Conveniently, informal credit involves very few stipulations on the way it is used. Farmers may use a group loan to purchase a dairy cow, but they may also use it to buy a new radio.

The flexibility and adaptability of informal, locally oriented forms of credit allow smallholders to be more adaptable when faced with urgent needs and other minor, or even major, crises. Addressing these issues surrounding rural credit is one way of ensuring the smallholder farming system is more resilient in terms of access to financial resources to strengthen livelihood activities which work together to support the entire household's food and nutrition security.

Endnotes

1 While K-Rep does offer individual loans, this particular branch is dedicated exclusively to "financial service associations" or group lending. The K-Rep Development Agency, the research and development arm of K-Rep, decided to pursue group lending in the late 1980s, encouraging group-based microlending as a way of expanding microenterprise in rural areas. K-Rep openly follows the Grameen Bank model based on the small-scale group lending schemes used among poor rural women in Bangladesh, which began in the mid-1980s. At K-Rep, clients may technically be male or female (though females are targeted since they would traditionally face more barriers to accessing bank loans) and are asked to form groups of five (called *watano* which is Kiswahili for five) before coming to K-Rep and asking for a loan. Depending on which loan program the group chooses (being a result of both group preference and the local manager's assessment of the group's needs), they may follow a shareholder system or a rotating credit scheme.

2 Kenya's Vision 2030 is a development plan covering the period 2008–2030. Among the many policies are provisions for greater employment among youth, gender equity, and economic growth in all sectors.

3 In the literature and in common development discourse, "unbanked" generally refers to anyone not yet incorporated into the formal financial sector.

4 Information accessed on March 4, 2013, on the Central Bank of Kenya website.

5 Ahead of the Destocking Crisis of 1939, a famous turning point in Kamba history, sisal plants were widely distributed as means of demarcating land in the Kenya (still in use today as a method for the demarcation of private land) and "hold the soil" in protection against erosion.

6 "Self-help group" is the broadest term both in the literature and in local use which suffices to encompass not only credit/savings groups, but also groups that exchange non-cash media.

7 In Kenya, "shylock" is a colloquial term used for anyone who lends money regularly and efficiently to a great number of people, often at a relatively high interest rate—in some ways, a "walking bank." http://www.standardmedia.co.ke/?articleID=2000106809 andstory_title=ease-access-to-credit-for-women Accessed April 24, 2014.

8 In Ukambani, land is customarily inherited from father to son after the son marries. From my experience in Makueni County, chiefs in Makueni District seem to be promoting the acquisition of title deeds (which aligns with the national agenda towards privatization of land), though farmers do not necessarily see it as a priority because of the added costs they incur. Women work the land just as much if not more than men, though are not considered owners of the land.

9 Officially, AFC is registered as a commercial "non-bank financial institution." AFC subsumed the Land and Agricultural Bank of Kenya in 1969 and took on the functions of a lending institution as a result. Prior to that, the AFC's mandate centered on ensuring the peaceful transfer of land to indigenous farmers and spurring development projects in rural areas (see the AFC website, last accessed on March 31, 2014: http://www.agrifinance.org/).

10 Reference: http://www.probonoaustralia.com.au/news/2012/09/world%E2%80%99s-poor-%E2%80%9Cunbanked%E2%80%9D-measuring-financial-inclusion

11 In recent years, the microfinance institutions K-Rep has found particular success with its relatively new "Financial Services Association" (FSA) arm. FSAs are separate institutions that lend only to what they refer to as "shareholder groups." Shareholder groups are self-organized groups. The FSA has various loan sizes which they offer to the group as a whole, but the group decides on how big a loan to take based on each member's estimated credit need. That is, each member ("shareholder") agrees to take a certain number of "shares" of the loan which the FSA gives to the group (e.g., if someone needs to plan an upcoming graduation celebration, perhaps that member will request ten shares, whereas someone with more "everyday" credit needs will only take one or two shares). Each shareholder's repayment capacity is assessed by K-Rep representatives to ensure that if someone does take, say, ten shares, she is poised to repay them in a timely manner. The first loan repayment cycle is always six months and if every member repays on time, the second cycle is one year, etc. The loan size is on a gradient such that the group loan can be increased in each cycle if the group prefers. The collateral required to be a part of these groups are minor assets only, and several group members visit each member's household to determine which assets they are comfortable with putting forth. It is the group members themselves who contractually sign each member's collateral away to the group.

12 Women, in reality, are known to do as much or more of the manual labour on the farm than men. However, the primary income-generating activity (and the profits thereof) of the household is generally managed by the man (Johnson, 2004).

13 Pots for cooking.

References

Akin, D. and Robbins, J. (Eds.)(1999). *Money and Modernity: State and Local Currencies in Melanesia.* Pittsburgh, PA: Univ. Pittsburgh Press.

Anyanzwa, J. (2012, October 8). CBK wants banks to deepen financial inclusion. *The Standard,* Nairobi, p. 26.

Ardener, S. (1964). The comparative study of rotating credit associations. *Journal of the Anthropological Institute of Great Britain and Ireland, 94*(2), 201–229.

Ardener, S. and Burman, S. (Eds.) (1996). *Money-go-rounds: The Importance of ROSCAs for Women.* Oxford, UK: Berg.

Bouman, F. J. A. (1995). Rotating and accumulating savings and credit associations: A development perspective. *World Development, 23*, 371–384.

Folke, C., Carpenter, S., Elmqvist, T., Gunderson, L., Holling, C. S., Walker, B. et al. (2002). *Resilience and Sustainable Development: Building Adaptive Capacity in a World of Transformations.* (Scientific background paper on resilience for the process of the World Summit on Sustainable Development on behalf of the Environmental Advisory Council to the Swedish Government). Stockholm, Sweden: Environmental Advisory Council, Ministry of the Environment.

Freeman, H. A., Ehui, S. K., and Jabbar, M. A. (1998). Credit constraints and smallholder dairy production in the East African highlands: Application of a switching regression model. *Agricultural Economics, 19*(1), 33–44.

Jaetzold, R., Schmidt, H., Hornetz, B., and Shisanya, C. (2006). *Farm Management Handbook of Kenya vol. II—Natural Conditions and Farm Management Information—*2nd edition part C East Kenya subpart c1 Eastern Province. Nairobi, Kenya: Ministry of Agriculture.

Johnson, S. (2013). Debt, over-indebtedness and wellbeing: An exploration. In I. Guerin, S. Morvant-Roux, and M. Villareal (Eds.). *Over-Indebtedness and Financial Inclusion* (pp. 64–85). Abingdon, UK: Routledge.

Johnson, S. (2004). Gender norms in financial markets: evidence from Kenya. *World Development, 32*, 1355–1374.

Karim, L. (2011). *Microfinance and Its Discontents: Women in Debt in Bangladesh.* Minneapolis: U. of Minnesota Press.

Kinyanjui, M. N. (2012). *Vyama, Institutions of Hope: Ordinary People's Market Coordination and Society Organisation.* Nairobi: Nsemia Publishers.

Meier, G. M. and Rauch, J.E. (2000). *Leading Issues in Economic Development.* New York, NY: Oxford University Press.

Ochieng, W. R. and Maxon, R.M. (Eds). (1992). *An Economic History of Kenya.* Nairobi, Kenya: East African Publishers.

Ouma, S., Njeru, S., Kamau, A., Khainga, D., and Kiriga, B. (2007). *Estimating the Size of the Underground Economy in Kenya.* (KIPPRA Discussion Paper, No. 82, Kenya Institute for Public Policy Research and Analysis Handbook). Washington, DC: World Bank.

Quisumbing, A. R. and McClafferty, B. (2006). *Using Gender Research in Development.* Washington, DC: IFPRI.

Rankin, K. (2001). Governing development: Neoliberalism, microcredit, and rational economic woman. *Economy and Society, 30*, 1, 18–37.

Robinson, M. S. (2001). *The Microfinance Revolution: Sustainable Finance for the Poor.* Vol. 1. World Bank Publications. Washington, DC: World Bank.

Schuler, S. R., Hashemi, S. M., Riley, A. P., and Akhter, S. (1996). Credit programs, patriarchy and men's violence against women in rural Bangladesh. *Social Science and Medicine 43*, 1729–1742.

Scott, J. C. (1998). *Seeing Like a State: How Certain Schemes to Improve the Human Condition Have Failed.* New Haven, CT: Yale University Press.

Shipton, P. (2010). *Credit between Cultures: Farmers, Financiers, and Misunderstanding in Africa.* New Haven, CT: Yale University Press.

Yunus, M. (1998). *Banker to the Poor.* India: Penguin Books.

7 Nested economies

Gendered small-livestock enterprise for household food security

Leigh Brownhill, Esther M. Njuguna, Erick Mungube, Malo Nzioka, and Esther Kihoro

Introduction

Kenya has an estimated chicken population of 32 million birds of which 81% (25 million) are counted as *indigenous chickens* (IC), which are kept in flocks of various sizes in over 90% of rural households (KNBS 2010). The term 'indigenous chicken,' in the Kenyan context, refers to a range of local types (*Gallus gallus domesticus*) that are distinguished from what are commonly known as 'exotic' breeds, such as Rhode Island Reds and Leghorns. These imported 'exotic' breeds are less well adapted to the free-range feeding system common in rural Kenya, and require more costly inputs for best production. The exotic birds are also referred to as 'layers' or 'broilers,' since the large-scale chicken industry produces these breeds for commercial egg and meat sales. We focus on subsistence-oriented production of backyard flocks of indigenous chickens, what Guèye (2000) calls "rural family poultry" systems. By subsistence-oriented production, we mean small-scale farm-level enterprises, which directly serve multiple household needs including nutritional security, socio-cultural requirements (such as gifts, donations, feasts), ecological services (pest control and manure), as well as income generation (sales of eggs and live birds at farm gate or in local markets).

Though we call the birds in our study "indigenous chickens" (IC), there are in fact thought to be few pure-bred indigenous chicken genotypes remaining in Kenya as of 2015. Several projects were implemented from the 1980s to the early 2000s with an intention of 'upgrading' indigenous chicken breeds through a cockerel exchange programme. Farmers were given an 'exotic' breeding cockerel in exchange for their indigenous ones. This program eroded the indigenous chicken genetic pool. A renewed appreciation by both farmers and scientists of the benefits of the indigenous chicken breeds (i.e., being well adapted to the local climate and environment, ability to scavenge and to brood) and their suitability within small-scale farming systems has led to efforts to recover and strengthen the local breeds through improved free-range management practices. A most recent example is the European Development Fund's Kenya Arid and Semi-Arid Lands project (2008–2011).

The research focus

In the KALRO-McGill food security research project, the main on-farm research focused on farmer-led evaluation of drought-resistant crop varieties and supportive natural resource management techniques. When the crop evaluations were well established after two seasons, the project's animal health research team polled participating farmers to seek their input with respect to the priority research areas that they judged would strengthen animal health within the small farming system. Farmers from the 54 participating farmer groups selected indigenous chicken management among their top priorities.

The small livestock component of the project began with the recognition that rural chicken production takes place in close relation to cropping systems. The drought-tolerant crops being evaluated by farmers were able to supplement chicken feed, and in turn, chicken manure could be used to provide organic fertilizer for crops. The mixed crop and small livestock practices, then, were mutually supportive, which offered us, as researchers, a window to investigate the dynamic and integrative character of different livelihood activities within the semi-arid farming system.

In eastern counties of Kenya, chickens are kept by most farmers, but the majority of birds are owned by women (see Figure 7.1). For this reason, when farmers in the project prioritized the IC livelihood enterprise and value chain as a preferred focus for the farmer-led research, it caught the attention of our project's gender team. This team subsequently joined with the animal health team to consider the social, nutritional, economic, and ecological dynamics of the enterprise. While the animal health team focused on in-field and in-class learning and evaluation with the project's 54 primary farmers' groups, the gender team added another dimension by bringing into the activities six women's farmer groups from Wote, Makueni County, who were otherwise not participating in the rest of the project's activities and research streams (called *non-participating farmers*). These groups' members were all farmers who raised chickens and expressed an interest in improving their enterprise to bolster their household food, nutrition, and income security. Because they were not part of most of the project's wider activities, the Wote groups' inclusion helped us to comparatively assess how improved indigenous chicken management could fit within the economies of 'typical' semi-arid farming households.

The purposive selection of these six additional women's groups was undertaken through snowball referrals, based on the contacts made between the gender researchers and women 'opinion leaders' in the Wote community. The selection took a turn, at one point, to become a two-way process. We had begun our research with focus group discussions involving, at first, five additional women's groups. In the course of these discussions, which took place over several days, a member of another women's group (Sisters of St. Joseph's) heard about the ongoing research and requested to be included in the study. The Sisters, who are Catholic nuns living in a church convent near Wote town, were subsequently added to the sample and the IC study activities. One member of the group in particular had grown up raising chickens with her mother, who sold eggs in bulk. This Sister was the main force behind the group's small-scale IC enterprise at the nunnery, and was very innovative

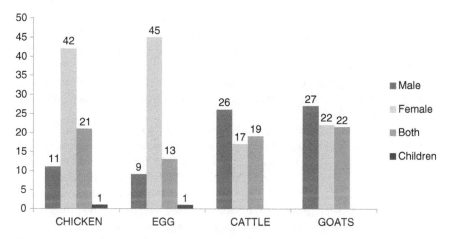

Figure 7.1 Household member(s) who owned the listed livestock.

with recycling materials to build stoops and separate housing and yards for chicks, brooding hens, and pullets. As nuns, the Sisters of St. Joseph were dedicated to service to the community. These characteristics added important dimensions to the researchers' understanding and to the outcomes for other farmers, who were ultimately influenced by the wide dissemination of information that these group members were able to mobilize.

Key questions that framed our study included: To what extent does indigenous chicken production contribute to the socio-ecological resilience of farming systems in the semi-arid midlands of Kenya? And in what ways does the enterprise contribute to (or undermine) the building of gender equity in the household, as the research activities helped improve production practices and raise income potentials? The income-generating and (youth) job-creating potentials of indigenous chicken small enterprises were also explored. These questions guided the research design and analysis. Grounded in the study's findings, we argue that by recognizing and strengthening the small-scale, subsistence-oriented common backyard chicken enterprises, the project contributed positively to multiple development goals, in particular improvement of household health and nutrition, facilitation of local marketing opportunities (for both women and men), and preservation of the semi-arid agro-ecology.

Methods and results

We addressed the research questions using four main data collection tools: (1) An IC farmer group survey assessed dynamics of the sector, including individual and farmer-group ownership, as well as issues covering the life cycle of the birds themselves, including health, housing, feeding, breeding, and marketing. Follow-up surveys were conducted among a smaller sample of farmers and results help support the chapter's main argument. (2) Focus Group Discussions assessed the contribution of subsistence-oriented backyard IC enterprises to household food sufficiency, income, and ecological well-being. Focus group discussions took place periodically, over the course of the project's tenure, at different seasons and stages of the project's activities. (3) Key Informant Interviews were carried out with farmers and public officials in Makueni, exploring questions of agricultural extension, markets, and policies. (4) And a gender survey examined power dynamics in the farming system, including issues of decision-making, access to and ownership of resources, and gendered divisions of labour.

We adopted an iterative "Triple-A" approach to data collection and analysis, that involved the three steps of Analysis, Action, and Assessment. This participatory methodology of knowledge creation, mobilization and integration was developed in the 1990s by researchers in UNICEF's Child Survival, Protection and Development program. UNICEF originally crafted this methodology for use in health and nutrition programs in Tanzania and Ethiopia. They acknowledged that farmers and other people are already engaged in *analysing* problems and challenges, taking *action* on the basis of that analysis, and *assessing* the outcomes in order to inform future action (Kavishe, 1995). What the approach added to the research process, then, was the opportunity to iteratively take stock of decisions, actions, and impacts as the research proceeded, and adjust direction if needed. The approach was subsequently adapted by feminists in the Tanzania Gender Networking Programme for use specifically in gender analyses in an African context (TGNP, 1993). We turn, next, to a presentation of the results of the research as they emerged in these three 'stages.'

Analysis of subsistence chicken production systems

Indigenous chickens in eastern Kenya are mainly owned by subsistence farmers, and are predominantly raised under the free-range system, as they have been for a very long time (Gichohi and Maina, 1992; King'ori et al., 2010). Rural farming households in Kenya have been identified as "custodian of these genetic resources," which have tremendous potential as part of resilient rural livelihood strategies (Magothe et al., 2012, p. 119). However, in the semi-arid regions in particular, the dry seasons bring with them a reduction in the availability of wild plants and a low insect population, which leads to slower bird growth and low productivity due to inadequate quantity and quality of available feed (King'ori et al., 2010). Slow growth rates, poor egg production (beginning only after 6 months of age), and poor reproductive performance (Ndegwa et al., 1996; Pedersen, 2002; Phiri et al., 2007) are compounded by high predation and high mortality rates (Mungube et al., 2008)—all of which undermine the potential of rural family poultry as an important subsistence livelihood contributor.

We examined indigenous chicken enterprises and farmers' practices to identify potential areas for improvement. Indigenous chicken production in the semi-arid midlands of eastern Kenya was found to be constrained by a number of problems including, most significantly, disease outbreaks, seasonal food shortages, inbreeding, and unhygienic and inadequate housing conditions (e.g., cold cement floors, lack of perches, lack of brooding spaces, insufficient ventilation). Additionally, insufficient access to market consumers of both eggs and chickens was reflected in the commonly expressed problem of the low prices offered by brokers and 'middlemen.' Newcastle Disease was found to be the biggest threat to flock survival, capable of wiping out entire flocks. There is no treatment except prevention through vaccination.

Despite the many constraints reported by the majority of participants, Kenya's farmers persist in keeping chickens, and periodically restart new flocks after partial or total losses. This persistence indicates farmers' almost ubiquitous determination to maintain chickens as part of the farm ecology; it is often said that a house is not a home without at least a few hens and a rooster in the backyard. In view of the constraints experienced by the majority of farmers, relatively simple and cost-effective measures were designed and introduced, and disseminated through peer-to-peer extension (see Chapter 2).

Research-for-development action: Training and service provision

To address the challenges that farmers faced in indigenous chicken production, the project team conducted a training program to cover the basics of improved chicken management, including low cost IC housing and biosecurity (e.g., regular cleaning and disinfection of chicken houses), feeding and nutrition, basic homemade ration formulation for feed supplementation, watering practices, egg handling, breeding and within-flock selection, predation, parasitism and disease management, business planning, and record keeping. In January and February 2013, KALRO held two one-week residential training sessions in Matuu, a town near the centre of the three counties participating in the INREF project. The training sessions were conducted by facilitators drawn from various disciplines, both from the public and private sectors, including two veterinarians, one agricultural economist, one social-anthropologist, and an animal nutritionist.

Members of the project's 54 primary farmer groups nominated one of their own to attend the training and become peer educators and service providers for their groups, to help educate and assist other group members to increase their poultry production. In addition to the representatives of the 54 farmer groups, nominees from seven other women's groups also participated in the training (these included representatives from the six women's groups from Wote, plus another Sister from a local Matuu nunnery). Emphasis was given to vaccination against Newcastle disease and supplemental feeding. Since feed can account for up to 75% of the variable costs of a poultry enterprise (Bell and Weaver, 2002), it is generally not feasible for cash-poor farmers to embrace supplementation using commercial feeds. Formulation of basic home-made rations for chicks, growers, and layers (adults) using locally available grains was therefore introduced as a more viable option for increasing farmers' access to supplemental feeds.

After the training, KALRO then provided the 54 farmer groups (and one of the additional women's groups) with starter flocks of 10 hens and one cockerel. The birds were of an 'improved' indigenous breed, bred at KALRO's National Animal Husbandry Research Centre at Naivasha. Each flock was maintained by one host farmer per group, whose farm then served as a site for the demonstration and evaluation of, and peer education on, the production techniques learned at the IC training session. Vaccination against common poultry diseases, especially Newcastle Disease, emerged as a key improvement in group members' poultry activities, leading to significant improvement in bird survival. The trained group members also served the needs of their groups by following up on the timely vaccination of group members' birds, distributing eggs (for incubation and reproduction), and advising group members on local production of supplemental feed (using traditional crop varieties) and low-cost housing (using local materials).

Assessment of outcomes

Within six months, the integrated IC management approach that farmers adopted in this project resulted in improved watering and feed supplementation, as well as a 65% increase in the uptake of vaccination against Newcastle disease. These improvements in turn led to better bird survival rates as well as improved egg production and chick survival rates. More birds meant that more eggs and meat (protein) were available to participating households. Women's groups were especially keen, and began to engage in group-based activities, such as purchase of chicks. One Wote woman farmer told us,

> *I personally was sent by the women's group to attend the Machakos Agricultural Show, to go and look for the stand where the Naivasha KARI improved chicken is, so that I may know what they are selling and then I can take the report to them. The group collected money from each member, and it is a group of 43 women, and pooled money so that each gets two chicks and takes them to her home.*

Feed supplementation was also improved, as by a farmer in Yatta:

> *After our service provider came back, he taught us about supplementing the chicken feed, I understood, for the first time, that chicken need a balanced diet and clean water, I realized then that I had been mistreating my chickens, I had to change* (Personal communication, 16 October 2013, Yatta).

This farmer changed the way she fed her birds: she used a hand miller to prepare a ration of three kgs maize, one kg of green grams, dolichoes, or beans, mixed together with dried green leaves. She also separated the birds according to their age groups. In a short time, she managed to increase her flock from 4 birds to 100.

Indigenous chickens not only helped to improve nutrition in the homes where they were raised. These chickens also "crossed the road" into very local value chains and market sales to improve especially women's household incomes. Better flock survival rates meant more sales of eggs and live birds—and a much greater market potential. The practices evaluated in the research worked to improve production significantly and generally overcame many production constraints. Market constraints, however, are often more difficult to breach than production constraints. Farmers in our study were more firmly in control of production than of marketing channels, thus our participatory work with farmers was biased in favour of their spheres of influence. We did make efforts with farmers to create marketing opportunity groups to facilitate aggregation and collective negotiations with buyers as well as to improve prices and spur collective innovation (see Chapter 10, this volume).

In 2014, follow-up discussions were held with the Wote women's groups to mobilize a group self-assessment of progress and achievements. Altogether, some 60 members representing six women's groups gathered to for a half-day learning event, at which break-out discussion groups addressed the diverse challenges they faced and innovations they had adopted. All groups reported improved feeding, watering, housing, bird survival, breeding, egg production, and marketing outcomes one year after the training and peer-to-peer education commenced. Not only did this learning event lead to sharing and exchange of ideas and innovations, but as the event came to a close, the group members decided (without the prompting or involvement of the researchers) to take their organizing efforts one step further. They took the opportunity afforded in being together to form a women's IC producers' association. Before the day was out, they had formed a board, held an election of officials, and set the date of their next meeting. They had also determined that the different women's groups would confer with their members to focus efforts on different links in the IC value chain, including egg production (for hatching, for eating), cockerel exchange, local feed production, organic manure composting, and marketing, whether at farm-gate, village market, or further afield.

Discussion: Nested economies

The positive outcomes of the project's indigenous chicken research activities need to be considered in their farming-system context. As previously mentioned, chicken farming in Eastern Kenya is a common practice, and an iconic part of the mixed, small-scale, and subsistence semi-arid farming systems. Our research activities were geared to fit within these local systems, and to strengthen the potential benefits from the articulation of backyard IC production, for both inputs (e.g., vaccines, feeds) and outputs (e.g., eggs and meat), as well as market and nutrition value chains. In light of the objectives of the INREF project, the household food security and nutritional benefits of increased IC production were paramount. As a result, the market links we envisioned and aimed to facilitate were in some ways subordinated to the nutritional goals. More specifically, we did not want to prioritize market production at the expense of household nutrition and farm agro-ecology, but rather we wanted to use market linkages, especially within very local value chains, to complement and support household food sufficiency, nutrition, and environmental protection objectives. So while marketing indigenous chicken products was also an objective, it was balanced by what we have called elsewhere 'non-priced values' of direct food supply and environmental services (Njuguna, Brownhill, Kihoro, Muhammad, and Hickey, in press). To emphasize the importance we place on the distinction between subsistence-oriented and market-oriented enterprises, we use the term 'nested economies' to describe the coexistence and inter-relations between the economies of household subsistence, the market, and the natural environment within which the first two are set (see Figure 7. 2).

This conception of nested economies can be distinguished from other similar models. Beugelsdijk (2009) focuses on "nested economies" as one of four types of 'social capital.' Social capital, or "network based processes that generate beneficial outcomes through norms of trust" (Durlauf and Fafchamps, 2004), is pertinent to the discussion in this chapter, since, in semi-arid farming systems in Kenya, group dynamics and collective action are highly valued productive assets and underlie the broader INREF approach (based on networked farmer groups and women's associations—see Chapter 2). The nested economies that Beugelsdijk describes as "embeddedness of individuals in trust-based micro networks, and the embedded-ness of these networks in overall society," seem to quite accurately describe the collective action observed in our research area, that is so critical to farmers' access to agricultural information (Guèye, 2009). However, we broaden the frame consid-erably to include not only social capital, but also the market-based and ecological realms within which such forms of social capital are equally relevant (e.g., through group-based marketing).

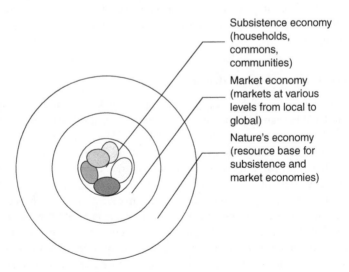

Subsistence economy (households, commons, communities)

Market economy (markets at various levels from local to global)

Nature's economy (resource base for subsistence and market economies)

Figure 7.2 Nested economies.

Dyer-Witheford's 'circuits of capital' (1999) takes a different perspective to denote four nested sites necessary for the generation of profit in the formal market economy; these include sites of production, consumption, social reproduction, and nature. His model describes how production of commodities is useful for the creation of profits only if those commodities are purchased and consumed. Consumption, in turn, is necessary to the processes of social reproduction, that is, the upbringing and education of new generations of producers and consumers. All of this activity is predicated on the use and exploitation of nature as a source of raw materials for commodity production (as well as subsistence production) and as a site of waste disposal. Dyer-Witheford (1999) identifies struggles between capital and the dispossessed in each realm: strikes plague the production circuit; consumers launch boycotts; students, women, and others struggle for their rights within the realm of social reproduction; indigenous people and environmentalists struggle to protect nature. These circuits, then, are not only elements of the circulation of capital, but also sites of gendered (and class) contention and change.

While we do not dispute the salience of Dyer-Witheford's conception, we see it as centred on a description of the capitalist economy. It is not a conception concerned with the distinctions between market-oriented and subsistence-oriented political economies. We take a different approach in our conception of nested economies—an approach grounded in research with peasant, small scale and subsistence farmers, whose daily livelihood pursuits are steeped more in subsistence than in market social relations (in comparison, for instance, to the producer and consumer in Dyer-Witheford's circuits of capital). This is not to say that we ignore capital, markets or value chains. We try to recognize class conflict by using the nested economies concept to analyse relations of contention and change between and among the three economies of subsistence, markets and nature. In the current case, we lay emphasis on the food and nutrition security impacts of value chains, and identify, through farmer evaluation and prioritization, livelihood practices (and social relations) that improve household food and nutrition security and the environment. From there we inquired into the process of linking farmers, and their prioritized enterprises, to markets that are accessible at locations, times, and scales suited to their socio-cultural and agro-ecological contexts.

The nested economies concept also helps frame the ways that we tried to avoid some of the gendered common pitfalls of agricultural development and research initiatives (Meinzen-Dick et al., 2014). In the past, many farmers would sell eggs and live chickens (or other agricultural products) instead of catering for the subsistence and dietary needs of their own households. One of our key objectives was therefore to raise on-farm consumption of own produce, to shorten the distance between families and the food and nutrients that they need. Therefore, in the project activities, the benefits of home consumption were emphasized and reinforced through formal nutrition education programs (FAO, 2013). In addition, often when eggs and chickens are sold, farmers may use the money to purchase foods which contain lower nutritional value (e.g., white bread, sugar) instead of diversifying the diet by purchasing more nutritious foods. This is not an uncommon practice in many contexts wherein cheap, high-calorie starches and empty carbohydrates are more accessible to cash-poor consumers than high-density nutritious foods. The project's nutrition education program therefore emphasized the importance and value of dietary diversity and nutritious food choices. The message reached participating farmers, one of whom told us,

> *I can now buy a dose of vaccine for my flock since the smallest pack takes care of 100 birds. I have enough eggs to feed my family, some of who are staying in Nairobi. Maybe I will be able to sell some eggs in the future, for now, I am meeting the needs of the family* (Personal communication, Yatta, October 2013).

We sought to avoid another common occurrence in research for development: that success in women's income-generating efforts leads men to move into these value chains and edge women out (Sorensen, 1996). Though a Kamba man interviewed in Wote in 2010 admitted that "In our tribe, we don't believe that a man…you cannot see a man carrying a hen to sell. No, no. That is the work of women"; not everyone feels the same way. In fact, across much of East Africa, both men and women can be found engaged in trade in both indigenous and exotic chicken meat and eggs. Njuki and Mburu (2013) show that not all women are as much in control over sales of chicken as the women are in Wote. Kamba women's apparent high degree of control over small livestock in Ukambani (in comparison to averages in the regions) deserves further study. For if there are specific socio-cultural norms at play which undergird Kamba women's power in relation to important and accessible livelihood resources, these norms might offer development researchers new insights for innovation in empowerment initiatives elsewhere.

A further question requiring further inquiry concerns apparent differences in patterns of ownership and control pertaining to indigenous versus exotic chickens. While free-range indigenous chickens are found in almost every rural home, the more input-intensive exotic birds have been introduced in Wote, as elsewhere in Kenya, through private or non-governmental agencies' projects, government programmes, and research initiatives. Thus a farmer's access to and control over exotic birds is in many ways tied to their participation in such projects. This in turn may produce particular relations of ownership and control that differ from farmers' ownership of ordinary backyard hens. For instance, in Njuki and Mburu's (2013) study of livestock ownership in Tanzania, half of women surveyed reported having full control over their exotic chickens, and could sell them without asking their husbands. The rest reported that they consulted with their husbands before selling; but none reported that their husbands had the right to sell the exotic birds, even if they consulted with their wives. This is in comparison to local indigenous chicken, where fewer than one-third of women reported selling without consulting their husbands. Almost one-fifth of women also reported that their husbands had rights to sell local chickens, mainly with consultation, but in some instances without it (Njuki and Mburu, 2013, p. 28).

Acknowledging the complexities and nuances of project-farmer interactions around small livestock research, we took steps in the project (e.g., through participatory market training) to address the value of women's control over income, as they are often more likely than men to invest in the household's food sufficiency (Bernasek, 2003; FAO, 2011, p. 44; Haddad, 1999; Quisumbing, and McClafferty, 2006; Thomas, 1997). A leader of a Wote women's group confirmed the improvement of women's food provisioning capacities through the project's efforts to improve backyard poultry enterprises:

> *We are selling many eggs, not like before. We are selling many small chicks to the nearby women groups and also the neighbours at large…. Whereby you can buy a [school] uniform for your child, you can buy books, you can meet your basic needs…. You can sell them, you can be sustainable and you cannot even borrow from a neighbour because right now you have the potential* (Interview, Wote, 2013).

The objective of the project was not to establish large-scale commercial chicken farming, which has been the context for much of the development of 'chicken improvement' technologies and management efforts, such as those focused on speeding production. Instead we draw lessons from these efforts for application within subsistence mixed farming systems to support household food and nutrition security, as well as ecological resilience at the farming-system level. Commercial chicken production is an input-intensive industry that can neither be accessed by the majority of small-scale farmers in our study area, nor could it offer the same widespread, direct nutritional, economic, and ecological benefits at the household level as small-scale IC enterprise can. While a few large industry players may dominate the chicken sub-sector, there are millions of small-scale and subsistence farmers who could benefit from the flourishing of backyard chicken enterprises, in which they maintain greater control over production processes, outputs, and benefits of trade. Farmers in Wote, for instance, recognized that small-scale and farmer-led enterprises keep decisions and direct benefits available to farmers' household:

> *We want to make a community-based organization. We will come together, sit down, and see how we can reach a bigger market; let's say Nairobi. That is our vision. After that, we make money, we feed our families, we educate them and we live a comfortable life* (Interview, Wote, 2013).

Small-scale farmers' cooperative marketing models illustrate that different niches or levels of markets can be occupied by different sellers, to satisfy diverse consumers' needs. The concept of nested economies highlights the importance of the re-establishment of very local markets (with short value chains) as well as multilevel marketing channels in which the few very large actors and many very small farmers play primary and complementary roles. Job creation and the potential for youth entrepreneurship is the next logical development of this 'nested' indigenous chicken sector. The self-employment of youth in a range of industries related to supporting small-scale farming enterprises, such as brick-making and carpentry for chicken coop construction, is a key link needed to strengthen very local IC value chains.

Conclusion

The indigenous chicken, which has a short production cycle and can be produced on a year-round basis, has the potential to contribute to 'smoothening' or evening out rural households' access to food and income across seasons, thus improving the constancy of food availability. As such, IC production has the potential to strengthen farmers' capacities to avoid food scarcities during the dry season. In addition, small-scale IC production can supplement incomes and improve dietary diversity, as well as provide inputs into crop production and related enterprises.

A pertinent political issue that makes this study both scientifically and practically important is the prioritization of indigenous chicken value chains at the level of new county agricultural directorates. Given the newly devolved governments' focus on improving IC enterprise in many of Kenya's 47 counties, the current study provides relevant research-backed recommendations which can readily fit into ongoing policy, extension, and development processes.

Both chickens and eggs are an established part of the household economy, often maintained by women as a general good and as a cultural and nutritional asset. Raising indigenous chicken is an accessible enterprise that responds well to effective methods for low-input, high-output very local value chain development (Njuki and Miller, 2013). And because it is already so widely adapted to conditions across the region and country, these positive impacts are possible to realize within a very short time in virtually every household that keeps small flocks of indigenous chickens.

Specifically, indigenous chicken enterprises have the potential to provide subsistence, market, and environmental benefits to women, men, and youth. We would suggest that these triple or nested benefits appear to be maximized when women's robust household economies are sustained, to serve as sources of direct subsistence and in mutually supportive relations with markets and the natural economies within which these activities are nested.

The perspective developed here suggests that future agricultural research, development, and policy will align with socio-ecological resilience objectives to the extent that they support the strengthening of the economies of subsistence and nature as much as they currently concentrate on market development. This support could come in the form of research funding priorities, development programming, and government budgets and policy frameworks. In practice, this means a realignment of the predominant market paradigm to put the market on par with both subsistence and ecological economies. Maintaining the gains observed in our study will ultimately depend on how well agricultural research, development, policy, and extension services embrace the necessity of strengthening the socio-ecological resilience of subsistence agriculture and nature's economy, as well as the effective articulation of the market economy in ways that reinforce relations of equity, inclusion, diversity, and farmer-led adaptation.

References

Bell, D.D. and Weaver Jr., W.D. (2002). *Commercial Chicken Meat and Egg Production* (5th ed.). Norwell, MA: Kluwer Academic Publishers.

Bernasek, A. (2003). Banking on social change: Grameen Bank lending to women. *International Journal of Politics, Culture and Society, 16*, 369–385.

Beugelsdijk, S. (2009). A multilevel approach to social capital. *International Studies of Management and Organization, 39*, 65–89.

Durlauf, S.N. and Fafchamps, M. (2004). *Social Capital* (Working paper series, No. 10485). Cambridge, MA: National Bureau of Economic Research (NBER).

Dyer-Witheford, N. (1999). *Cyber-Marx: Cycles and Circuits of Struggle in High-Technology Capitalism*. Urbana and Chicago: University of Illinois Press.

FAO. (2011). *The State of Food and Agriculture: Women in Agriculture, Closing the Gender Gap for Development*. Rome, Italy: FAO.

FAO. (2013). *The State of Food and Agriculture, 2013: Food Systems for Better Nutrition*. Rome, Italy: FAO.

Gichohi, C.M. and Maina, J.G. (1992). *Poultry Production and Marketing* (Paper presented in Nairobi-Kenya, November 23–27, 1992). Nairobi, Kenya: Ministry of Livestock Production.

Guèye, E. (2000). The role of family poultry in poverty alleviation, food security and the promotion of gender equality in rural Africa. *Outlook on Agriculture, 29*(2), 129–136.

Guèye, E. (2009). Small-scale family poultry production: The role of networks in information dissemination to family poultry farmers. *World's Poultry Science Journal, 65*, 115–124.

Haddad, L. (1999). The earned income by women: Impacts on welfare outcomes. *Agricultural Economics, 20*(2), 135–141.

Kavishe, F.P. (1995). Investing in nutrition at the national level: An African perspective. *Proceedings of the Nutrition Society, 54*, 367-378.

King'ori, A.M., Wachira, A.M., and Tuitoek, J.K. (2010). Indigenous chicken production in Kenya: A review. *International Journal of Poultry Science, 9*, 309–316.

Kenya National Bureau of Statistics. (2010). *Livestock Census Data*. Nairobi: KNBS.

Magothe, T.M., Okeno, T.O., Muhuyi, W.B., and Kahi, A.K. (2012). Indigenous chicken production in Kenya: Current status. *World's Poultry Science Journal, 68*, 119–132.

Meinzen-Dick, R., Quisumbing, A.R., and Behrman, J.A. (2014). A System that delivers: Integrating gender into agricultural research, development, and extension. In A.R. Quisumbing, R. Meinzen-Dick, T.L. Raney, A. Croppenstedt, J.A. Behrman, and A. Peterman (Eds.), *Gender in Agriculture: Closing the Knowledge Gap* (pp. 373–391). Netherlands: Springer, Rome: FAO.

Mungube, E., Bauni, S.M., Tenhagen, B.-A., Wamae, L., Nzioka, S.M., Muhammed, L., and Nginyi, J. (2008). Prevalence of parasites of the local scavenging chickens in a selected semi-arid zone of Eastern Kenya. *Tropical Animal Health and Production, 40*, 101–109.

Ndegwa, J.M., Kimani, C.W., Siamba, D.N., Ngugi, C.N., and Mburu, B.M. (1996). *On-Farm Evaluation of Improved Management Practices on Production Performance of Indigenous Chickens* (October–December 1996 quarterly report). Naivasha: National Animal Husbandry Research Centre, Kenya Agricultural Research Institute.

Njuguna, E.M., Brownhill, L., Kihoro, E., Muhammad, L.W., and Hickey, G.M. (in press). Gendered technology adoption and household food security in semi-arid Eastern Kenya. In J. Parkins, J. Njuki and A. Kaler (Eds.), *Towards a Transformative Approach to Gender and Food Security in Low-Income Countries*. London, UK: Earthscan.

Njuki, J. and Mburu, S. (2013). Gender and ownership of livestock assets. In J. Njuki, and P. C. Sanginga (Eds.), *Women, Livestock Ownership and Markets: Bridging the Gender Gap in Eastern and Southern Africa*, (pp. 21–38). London, UK: Earthscan.

Njuki, J. and Miller, B. (2013). Making livestock research and development programs and policies more gender responsive. In J. Njuki, and P. C. Sanginga (Eds.), *Women, Livestock Ownership and Markets: Bridging the Gender Gap in Eastern and Southern Africa*, (pp. 111–128). London, UK: Earthscan.

Pedersen, C.V. (2002). *Production of Semi-Scavenging Chickens in Zimbabwe* (PhD thesis). The Royal Veterinary and Agricultural University, Copenhagen, Denmark.

Phiri, I., Phiri, A., Ziela, M., Chota, A., Masuku, M., and Monrad, J. (2007). Prevalence and distribution of gastrointestinal helminths and their effects on weight gain in free-range chickens in Central Zambia. *Tropical Animal Health and Production, 39*, 309–315.

Quisumbing, A.R. and McClafferty, B. (2006). *Using Gender Research in Development*. Washington, DC: International Food Policy Research Institute.

Sorensen, P. (1996). Commercialization of food crops in Busoga, Uganda, and the renegotiation of gender. *Gender and Society, 10*, 608–628.

Tanzania Gender Networking Programme. (1993). *Gender Profile of Tanzania*. Dar es Salaam: TGNP.

Thomas, D. (1997). Incomes, expenditures and health outcomes: Evidence on intrahousehold resource allocation. In L. Haddad, J. Hoddinott, and H. Alderman (Eds). *Intrahousehold Resource Allocation in Developing Countries*. Baltimore: Johns Hopkins University Press.

8 Women second

Reflecting on gendered resilience within formal regulatory policies for forest-based livelihood activities in Kenya

Stephanie Shumsky, Kimberly L. Bothi, Elizabeth Nambiro, and Patrick Maundu

Introduction

Forests enhance socio-ecological resilience through a variety of mechanisms: supporting and regulating the natural environment (e.g., watershed management and carbon sequestration), providing an abundance of wild goods from fuel to medicine and providing cultural spaces and resources for communities (e.g., for important ceremonies). In this chapter, we will limit our discussion to a subset of these ecosystem services which has traditionally received the most coverage in academic literature, public policy and private industry—nature's provisioning services (Bennett, Peterson, and Gordon, 2009; Boyd and Banzhaf, 2007; Kalame, 2011; United Nations 2005). Fuels, foods and fodder produced within forest ecosystems as well as other resources located within them, like fresh water and arable land, are highly valuable from environmental, economic and social perspectives. Many components of resilience are therefore well represented within the forest, in particular the natural resources that can be harvested and used to diversify livelihoods, improve household nutrition and support ecological balance.

Here, we explore issues of access that either support or prohibit stakeholders' engagement in decisions around local natural resource management. Our conceptualization of resilience also focuses on the ability of systems to adapt in the face of change by preparing for upcoming shocks and being flexible enough to accept uncertainty. For us, this means that both forest users and forest policymakers may need to be more accepting of changes, especially those which extend the nutritional and ecological benefits of forests to more members of households and communities. Forests provide essential hidden harvests and hungry-season foods, in addition to alternative livelihoods and income-generating opportunities during times of food insecurity and economic hardship for many subsistence-farming households (Vira, Wildburger, and Mansourian, 2015). Recognizing the importance of these resources to many vulnerable households, we focus on the ways in which existing institutions and forest policies in Kenya may be gender biased, with implications for household food security and resilience. We then identify some ways in which more equitable forest policy-making and enforcement may be conceived in order to better promote gendered resilience.

According to the Kenya Forest Service (KFS) 2009/10–2013/14 strategic plan (2009), over one million households living within five kilometres of forests depend on these ecosystems for a variety of products (Box 8.1) and services, including "cultivation, grazing, fishing, water and other benefits." Researchers have long recognised the contribution of forest produce to food baskets (Burlingame, 2000; Scoones, Melnyk, and Pretty, 1992), household economies (Arnold and Perez, 2001; Zulu, 2010) and medical care (Iwu, 2014), as well as the disproportionate labour allocation by women in the collection, transformation and sale of forest products like honey, fodder and medicine (Byron and Arnold, 1999). Women's management of these forest products contributes to a variety of the resilience characteristics identified in Chapter 1, from diversifying incomes and nutrition sources (Ellis and Allison, 2004) to acting as a bank for community traditions, shared knowledge and understanding system dynamics (Folke, 2004). Despite the resilience-enhancing character of women's subsistence forestry in many developing area contexts, there often remains a gendered bias in practice, with women performing much of the labour involved in the use of forest food products, while men often retain greater access to and control of forest resources due to traditional land tenure arrangements, gender identities within households (Coulibaly-Lingani, Tigabu, Savadogo, Oden, and Ouadba, 2009) and access to extraction technologies (Veuthey and Gerber, 2010).

Within the rather significant body of literature on resilience, natural resources and the environment that considers issues of gender, women tend to be painted as especially vulnerable, possibly culpable and at the same time comparatively virtuous actors (Turner and Brownhill, 2006). The rise of ecofeminism, based on an idea that women work closely with nature, resulted in the mobilization of women to accomplish environmental projects, such as the hugely successful Green Belt Movement in Kenya (Maathai, 2004). Social movements, as well as academic debates, subsequently influenced strategic interests in the international development community. At times, however, development agencies' gender and environment projects have been undertaken without considering women's ability to access the necessary resources and succeed in these initiatives and without considering their time and energy spent for little tangible benefit (Leach, 2007).

Similar comparisons continue to be made in research on natural resource management, for example in Cameroon, where women's more diffuse and consistent mangrove harvesting practices were described as both undermining ecosystem sustainability while also playing a role in conservation and contributing

---

Box 8.1 Defining forest produce.

Kenya Forests Act (2005) defines "forest produce" as "bark, bat droppings, beeswax, canes, charcoal, creepers, earth, fibrewood, frankincense, fruit galls, grass, gum, honey, leaves, flower, limestone, moss, murram, myrrh, peat, plants, reeds, resin, rushes, rubber, sap, seeds, spices, stones, timber, trees, water, waxwithies and such other things as may be declared by the Minister to be forest produce for the purpose of this Act."

to degradation only as a consequence of men's initial harvesting activities (Feka, Mazano, and Dahdouh-Guebas 2011). Such complex and contradictory outcomes have led scientists considering gendered natural resource management to begin to frame issues from resource access perspectives and more broadly consider disenfranchised groups and institutional shortcomings (Leach, 2007). The research reported in this chapter seeks to bridge the gap raised by Leach by considering environmental policies through a gendered lens without oversimplifying nuances that exist within and across communities, and by acknowledging women as individuals with agency. In an attempt to avoid the trap of relegating women to the roles of victims, culprits or saviours, this chapter will consider the current segregation of forest-based harvesting activities and related policies, and its impact on household food security. We do so without making overt value judgments on the conservation implications of these activities, but rather leaving the readers to consider the context and likely outcomes of specific harvest practices for themselves.

Formal forest policies play a significant role in determining which demographic groups can access forest resources for subsistence use, commercial transformation and cultural applications. Governments in the global south have transferred large amounts of land from traditional tribal or community ownership to private entities and public institutions, effectively minimizing the land area under common property management (Monbiot, 1994). As of the most recent Food and Agricultural Organization of the United Nations (FAO) forest resources assessment (FAO, 2010), 80% of global forests were publicly owned, with African countries posting the highest percentage of public ownership worldwide. The new Kenyan constitution promulgated in August 2010 included environmental provisions in three of the 18 chapters and two of the six schedules, with a clear presentation of the constitutional right to a clean and healthy environment in Article 42 (Republic of Kenya, 2010). These provisions have been identified as increasing the opportunities available to improve environmental management in Kenya and providing a clear entry point for citizens and other stakeholders to include environmental considerations in the country's continuing development (Mwenda and Kibutu, 2012). The constitution restates the status quo of public control of forested lands in Kenya, where the vast majority of timberlands have been converted to publicly owned conservation areas through Section 20 of Kenya's Forests Act (2005). This Act claims all forests—aside from those owned by private actors or local authorities—as property of the state (Matiru, 2000). The registration of forests, as outlined in the Environmental Management and Coordination Act (2000), also falls under national jurisdiction and even applies to private landowners.

Overall, 97.8% (3.5 Mha) of Kenya's forest areas are owned by the Kenyan government, while 89.9% of other wooded lands (31.6 Mha) are managed by the state, leaving a total of only 3.5 Mha under private control (FAO, 2005). About 20% of the forest area controlled by the government has been set aside as national parks and reserves, or managed strictly for biodiversity conservation. Here, we concentrate our analysis on the State forests, which the constitution (Sec. 34.1) allows to be used for various purposes including apiculture, silviculture and infrastructural development. The Kenya Forest Service (KFS) is in charge of monitoring these activities, setting development priorities, and enforcing regulatory policies, which are detailed in periodic strategic plans (e.g., KFS, 2009) and other government documents. The new constitution (Republic of Kenya, 2010) enshrines the ecological objective of maintaining 10% tree cover on a national basis, while also calling for the protection of traditional interests and access to those same lands. These provisions create both the opportunity for the advancement of women's rights of access to forests for subsistence uses and at the same time delineate a potential site of conflict across stakeholders and the various levels of policymakers, enforcement agents and other state actors. Sihanya (2011) details the potential challenges due to misinterpretation of the constitution itself, intentional delays, confusion about implementation responsibilities within the bureaucracy and a lack of specific implementation pathways to comply with complex requirements (e.g., the rule limiting representation of any public body to two-thirds of the same gender) in a comprehensive analysis of the implementation process.

Ensuring equitable access to forest resources for both women and men, eliminating discrimination and promoting gender-aware programming and interventions are necessary to ensure sustainable forest management, as well as to fulfil important ethical considerations (FAO, 2012). Historically, men have dominated collective institutions and policy-making bodies (Pandolfelli, Dohrn, and Meinzen-Dick, 2007) and continue to do so despite support for increasing women's participation from a variety of public and private stakeholders (Nzengya and Aggarwal, 2013). Women's lack of membership on regulatory bodies has contributed to inequities in policy and access, which are compounded by passive behaviours sometimes associated with women who do take part in regulatory processes, such as limited contributions to discourse and the low regard that others often have for these con-sultations (Agarwal, 2001). Gender-biased policies often leave women burdened with greater costs (labour, time, economic and personal safety) to access forest resources. Women tend to enjoy lower profits from benefits-sharing schemes and face other consequences brought on by the sidelining of issues considered to be women's domains, like fuelwood scarcity on a local scale (*ibid.*). Disparities in legal protections and rights, such as inheritance laws that exclude women from property ownership, are also responsible for unbalanced power relations between genders (Meinzen-Dick, Brown, Feldstein, and Quisumbing, 1997). These inequalities affect intra-household decision-making and can have significant social and economic consequences (Quisumbing, 2003). Therefore, a more nuanced gendered analysis stands to help improve formal forestry institutions and policies so that they can better support more equitable access to resources in the future (Rocheleau and Edmunds, 1997).

Research objective

This research seeks to analyse both (i) regulations on access to natural resources, as delineated in the Kenya Forests Act (2005), and (ii) the enforcement of those rules by the Kenya Forest Service using the gendered resilience framework described in Chapter 1. We divide our chapter into three sections to illustrate the differences in policy and enforcement across gender, resource type and marketability. The first section focuses on subsistence use of wild foods, medicinal plants, honey, water and firewood; the second section describes market-oriented harvests like charcoal production, larger amounts of firewood, non-timber forest products (NTFPs) and commercial water reservoirs; and the final section examines land-related usage, such as grazing rights and farmland rental (Figure 8.1).

It is important to note, however, that traditionally gendered activities are not synonymous with absolute gender-based divisions of labour in this case, nor are the policy and practical biases we observed applicable to all the utilization of natural resources associated with women. In a multiple case study analysis, Sunderland et al. (2014) found that gender roles in natural resource use vary widely, depend on the specific conditions in a given area, and often do not conform to Western interpretations of sole and exclusive land ownership and usage rights. Many forest-based livelihood activities are not specifically segregated by gender and each day brings a continuing evolution of women's legal, cultural and *de facto* rights, which further complicates these issues. Through our analysis in this chapter, we hope to enhance understanding of gendered issues of forest access, and to draw attention to the potentially harmful biases that exist in forest policies, which may undermine women's food-provisioning capabilities.

While it is tempting to oversimplify differential access conditions using a single definition of what constitutes a 'user community,' it was important to us to apply multiple perspectives to more fully understand the nuances of these complex relationships. Conflicts between ethnic groups over unequal access to resources have long been blamed on natural resource scarcity (Castro and Neilsen, 2003). However, recent research suggests the more likely driver of most resource conflict is a combination of historical social, political and economic factors, exacerbated by weak formal institutions and biased policies at the local level (Opiyo, Wasonga, Schilling, and Mureithi, 2012). Moreover, Haro, Doyo, and McPeak (2005) detail different natural resource management practices and access customs across various ethnic groups in Kenya, and ultimately find that a common framework of household-level decision-making and hierarchical authority structures is shared by many. This research therefore seeks to examine the effect of formal policies on natural resource access through a gendered lens, but also acknowledges the numerous other factors and classifications that also have a role in shaping differential access to forest resources in Kenya.

Section I: Subsistence collection of natural resources from the forest

Many researchers have attempted to identify and quantify the products collected from forests for household use, for example, by applying economic valuations (Campbell and Luckert, 2002; Delang, 2006; Dovie, Witkowski, and Shackleton, 2005; Emerton, 1996) or by gathering data through participatory activities (Robinson and Lokina, 2011; Termote, 2011). A review of recent literature shows

Figure 8.1 Examples of forest-based livelihood products in rural Kenya.

the general consensus of these investigations, which suggests that households in developing area contexts rely heavily on forests to

1   Support food and nutrition security through the collection of wild edible plants (Maundu, Ngugi, and Kabuye, 1999; Shumsky, Hickey, Pelletier, and Johns, 2014a);
2   Provide energy from fuelwood and charcoal (Mahiri and Howorth, 2001) including for nearly 30 million people in Kenya (approximately 71% of the population) who use wood fuel to cook their meals, and supply an estimated 55.1% of all primary energy in the country for meal preparation, heating and boiling water (FAO, 2014);
3   Sustain access to other NTFPs like medicinal plants (Kokwaro, 2009; Njoroge, Bussmann, Gemmill, Newton, and Ngumi, 2004) and honey (Crafter, Awimbo, and Broekhoven, 1997; Muli, Munguti, and Raina, 2007), which are also collected from forests and conservation areas, and contribute to the nutrition and general well-being of household members;
4   Maintain water availability for household use, livestock and agricultural irrigation since forest springs (natural and improved) can provide a clean and fairly consistent source of water, especially during dry months where seasonal rivers and shallow wells are less reliable (Emerton, 1996); and
5   Provide construction materials like poles and timber for house building, which is considered the most economically valuable livelihood activity for forest user groups (Kabubo-Mariara, 2013).

When it comes to forest-based activities in semi-arid Kenya, divisions of labour by gender are often fairly rigid. Traditional labour distribution often leaves women to accomplish more time-consuming subsistence tasks like fuelwood and water collection (Helander and Larsson, 2014; Were, Swallow, and Roy, 2006), as well as the harvest of wild foods and medicinal plants (Harris and Mohammed, 2003). Men and boys within the household tend to take on harvests of larger quantities of firewood or produce destined for markets (discussed in Section II), and activities perceived as dangerous, such as honey collection (Eriksen, Brown, and Kelly, 2005) and picking wild fruits from tall *Tamarind indica* and baobab trees (Shumsky, field observation). All other activities related to wood, such as timber harvests for construction and sale, are predominately carried out by men (Cavendish, 2000). These trends in labour distribution are fairly consistent across communities throughout sub-Saharan Africa (Brown and Lapuyade, 2001; Scoones et al., 1992). This highlights the potential relevance of our analysis to diverse geographical landscapes as national forest policies continue to evolve in support of sustainable development objectives and forest user rights internationally.

The Kenya Forests Act (2005) includes several provisions that apply to the subsistence-level collection of forest produce and access to natural resources. There are some valuable inclusions, such as definitions of forest communities (Part 1—Preliminary), protections for 'traditional user rights' (Sec. 21), and several references to sustainable production of wood and non-wood products (Preamble,

Sec.6, Sec.40). However, the Act goes on to place restrictions on access to forests classified as natural reserves (Sec. 31), to require formal management plans for any entry into indigenous forests (Sec. 40) and to outline a complex and costly procedure for community-group participation in certain activities and programmes that includes requirements for a group constitution, accounting records, and environmental management plans to monitor and conserve biodiversity (Sec. 45). Finally, Part V of the Act covers enforcement and details specific activities that are prohibited unless a license is obtained (Sec. 52), such as harvesting forest produce or collecting honey, and lists the penalties for breaking these rules (Sec. 55, 57).

Under Section 59(b) of the Forests Act (2005),[1] the KFS is granted rights to set prices and procedures for all harvest and use activities that take place in forests. While it is interesting that a permit for filming movies (40,000 KSh per day) and the right to harvest a Christmas tree (300–700 KSh, depending on height) are detailed, the document conspicuously overlooks household-level collection of most forest produce. Wild foods and medicinal plants are not mentioned at all, while water collection is discussed only in large quantities (e.g., tanks with a 6 m diameter) and requires substantial licensing fees (5,000–1,000,000 KSh). Rural farming families living on the edge of forests are not the target audience for the 2012 Forest Rules. Furthermore, male-dominated activities are more clearly defined and regulated, such as the permit prices to cut down timber for construction poles or charcoal production—these values are detailed to the species and tree diameter. The trend, however, does not apply across the board, as the rights to collect honey are available only for community user groups, which is not an easy classification to obtain, and at 1500 KSh a year, not really accessible for most rural people in semi-arid Kenya. Another key exception to the lack of focus on subsistence use is the monthly fuel license (MFL) for only 100 KSh, which allows for the collection of one head-load of firewood per day over a 30-day period.

Section II: Harvest of forest produce for income generation

Income generation from forests is a common topic of research, policy and regulation. For instance, the collection of wood and NTFPs for sale, construction and charcoal burning can help bridge seasonal gaps in income, provide alternative livelihoods during times of crop failure (Zulu, 2010) and reduce income inequality within communities (Fisher, 2004; Mamo, Sjaastad, and Vedeld, 2007). The low entry barriers for such activities, such as minimal start-up costs and lack of specialised tools, in addition to the cultural acceptance of participation by the rural poor, makes trade in natural products an attractive economic strategy, especially for vulnerable groups such as women who typically engage more in informal markets (Shackleton and Shackleton, 2004). Despite several studies indicating the difficulty of balancing conservation and development objectives to sustainably support both biodiversity and economic growth (Kusters, Achdiawan, Belcher, and Ruiz Pérez, 2006; Ros-Tonen and Wiersum, 2005), governments, NGOs and foresters in the field continue to push communities toward commercial activities and value-added export chains for profit. In Kenya, an average of 40% of non-commercial providers of agricultural extension (both government and private NGOs) included value addition education and training in their activities with farmers (Muyanga and Jayne, 2006). Some examples of value-added forest products destined for markets include edibles such as juices, concentrates and candies made by processing wild fruits and honey, as well as cosmetics (KEFRI website) and herbal medicines (Mukonyi and Lwande, 2007).

As mentioned, women tend to be more involved in forest resource harvest and, more generally, forest management than men, especially where NTFPs are concerned. The importance of forest-based income for women has also been highlighted in other studies (see Ajonina et al., 2005; Mbuvi and Boon, 2009). These underline how heavily this demographic group relies on the returns from their alternative livelihood activities to cover household education, health and nutrition expenses (Adedayo, Oyun, and Kadeba, 2010). However, when greater economic benefits enter the picture, divisions in access and benefits start to shift. Using a global dataset, Sunderland et al. (2014) found that men tend to be responsible for collecting forest products with income generation in mind, and tend to gain a larger profit from those activities than women. This may be due to men's improved access to technologies, like farm machinery, that enhance their ability to exploit natural resources more effectively and at a larger scale (Veuthey and Gerber, 2010). Another possible explanation is the ability of men to dedicate time and energy to such activities over a sustained period. Women, on the other hand, are unable to work continuously due to other domestic obligations (Eriksen et al., 2005), and are often forced to make up agricultural labour deficits that occur when men pursue these typically off-farm activities (Zulu and Richardson, 2013). Overall, women tend to be displaced from forest-based value chains as they become more profit-able (Shackleton, Paumgarten, Kassa, Husselman, and Zida, 2011) and as increased mechanization allows for upscaling and a shifting focus to export markets (Hasalkar and Jadhav, 2004).

In addition to these considerations, Shackleton et al. (2011) describe how culture, traditions and logistics often combine to limit women's access to markets, negating their ability to optimise profits (see also Neumann and Hirsch, 2000). In a case study from Burkina Faso, Shackleton et al. (2011) connected less lucrative NTFP trade outcomes for women to both the social stigma that prevents them from using bicycles to transport their produce to larger markets and the religious considerations that discourage women from travelling far from home to connect with actors higher up the value chain. In Kenya, traditional perspectives often delineate forestry and tree products (i.e., timber, poles, charcoal and construction materials) as male-dominated fields (Rocheleau and Edmunds, 1997), while current discourse from policymakers and profit-driven commercial interests on timber harvesting perpetuate that legacy (Mai, Mwangi, and Wan, 2011).

Small-scale commercial activities like the ones mentioned above are clearly defined and regulated in the Kenya Forests Act, and the fees and charges supplement that law (KFS, 2012). Many of the income-generating activities primarily undertaken by women, such as the sale of wild foods and medicinal herbs, require licensing fees and community forest association group structures. The costs generally fluctuate based on what the on-site forester estimates market prices to be, making sellers quite vulnerable to price volatility through informal arrangements. However, some of the forest products traditionally exploited by women for market-based motivations are not as difficult to access (Box 8.2), including withies (vines used to weave baskets), which are priced at just 10 KSh ($0.12 CAD) per piece.

As previously mentioned, men tend to play a dominant role in the tree-based resource activities which are explicitly defined in the formal forest documentation and legislation of Kenya. Prices for livestock fodder by the bag, salvaged firewood by the cubic metre (a large quantity often intended for resale) and other products for sale are fixed, fairly affordable (50 and 600 KSh respectively) and even reduced in some cases due to community demands for lower fees (personal communication, Machakos Forester, June 4, 2014). Stumpage fees—the prices for trees felled for use as construction materials like poles or fuel by way of charcoal burning—are also fixed by tree species and trunk diameter or quantity, with price reductions when foresters allow for thinning of timber stands to prevent overcrowding and maximise yields. While community forest association group members may be able to take better advantage of positive forester-community relationships to receive preferred access to timber resources, these formal community-based organizations are not required to access timber resources.

Box 8.2 Luck of the draw on enforcement.

A wide spectrum of policy interpretation and enforcement exists across diverse geographic and socio-economic scales in Kenya. The implementation of participatory forest management outlined in the 2005 law has been inconsistent, leaving some communities to navigate the complex web of permits and punishments on their own. Other communities have been sensitized about the new regulations and supported through the transition, including most of the high-potential areas, or, more specifically, communities that are home to substantial resources, experience conditions more conducive to agricultural production and, generally, hold greater wealth and access to resources than other marginalized communities.

Community forest associations that have made it through the registration process manage forests with their KFS counterparts and orchestrate compromises like open access to water, medicinal herbs and other products for subsistence use without permit fees or restrictions. These groups benefit from enticing trade-offs, such as gained timber rights and access to construction materials (for a fee) in exchange for informing on neighbours' illegal activities and prioritizing shared conservation goals. In areas that haven't transitioned to participatory forest management, a lack of community forest associations and hostile KFS-community relations create a very different situation, where household collection of wild foods and other products is closely (and inconsistently) regulated, expensively permitted and harshly punished.

Section III: Access to land located within state forests

Our final section deals with other types of natural resource use in forests, namely the ability to rent land for farming or grazing livestock. These are particularly interesting cases, as they often defy formal definitions of land tenure and require their own considerations as a consequence of the historical reliance on common property resources or unowned "bush" lands in semi-arid Kenya (Rocheleau and Edmunds, 1997). As a father's land is subdivided again and again, descendants are often left with smaller holdings and ever-dwindling soil fertility to harvest their crops, graze their livestock and eke out a living (Murton, 1999). This is especially true in the arid and semi-arid lands (ASALs), which make up 80% of Kenya's land mass. Across the country's ASALs, the majority of farmers diversify their food production and income generation strategies in order to survive the stresses of an uncertain climate, limited access to capital and poor local infrastructure, all of which prevent effective and sustainable local market engagement. Livestock and agroforestry are therefore useful supplementary livelihood strategies that take advantage of ecosystems that are well adapted to dry conditions and more traditional livelihood practices. As of 2015, renting land to farm from state-controlled forests (Box 8.3) is illegal throughout

---

Box 8.3 Kenya's shamba system.

The 'shamba' system, also know as 'taungya,' was introduced to Kenya from Southeast Asia in the early 20th century by the British colonial government. In theory, the eco-friendly agroforestry systems of food crops and tree saplings were a welcome idea, allowing the Kenya forest service (KFS) to add forest cover at low cost, and providing landless farmers a chance to rent cheap yet fertile farmland to feed their families and generate income. As the saplings grew to shade crops over 3 years or so, farmers were expected to move to another plot.

However, support for the shambas has since evaporated due to corruption and environmental degradation. Politicians and local foresters abused the system to claim huge acreages for themselves and solicit bribes, while outsiders pushed out local communities to gain lands far from their ancestral homes. Farmers themselves even resorted to illegal logging and hindering sapling growth to avoid moving plots.

In 1985, all forest and leasing agreements were banned with limited success due to low enforcement. The concept was revived in 1993 under a new name—non-resident cultivators (NRCs)—followed by a new ban in 2002. Since 2008, the system has existed in a few areas as a pilot program under another new name, PELIS, or, the "plantation establishment and livelihood improvement scheme." Early data suggest that income from forest crops represents the largest portion of all earnings for NRC households, especially for those without owned land

Proponents of the shamba system cite low reforestation costs, reductions in landless squatters, increased food security and improved opportunities to address urban flight and youth unemployment; however, history has demonstrated that the risks of political corruption, tribal tensions and forest degradation may accompany and outweigh such benefits (Kabubo-Mariara, 2013; Witcomb and Dorward, 2009; Yatich et al., 2007).

most of Kenya, but remains an important consideration, as it may one day return as a common practice if local media reports prove correct (Njoroge 2011).

Livestock-rearing is a common option for livelihood diversification in the ASALs, producing some 95% of all household income in the region and employing 90% of the ASAL workforce (50% of the country as a whole). A 2012 Procasur report on the livestock sub-sector in Kenya stated that keeping cattle, goats, sheep and/or camels was the predominant source of income for one quarter of the national population who identify as pastoralists. Besides pure income generation, livestock-rearing provides direct-use benefits like credit, risk sharing/group insurance, transport, agricultural traction and fertilisers, not to mention food products like milk and meat for sale or household consumption (Behnke and Muthami, 2011). Livestock has also been shown to augment social status as a demonstration of household wealth and respect for cultural traditions (Ouma, Obare, and Staal, 2003).

These benefits, however, are not without substantial input costs, including water, feed and veterinary care, which pose especially large burdens on smallholder farmers who cannot afford the inputs to augment their management practices. Commercially available feed is sometimes difficult to access and subject to price increases due to market fluctuations and other outside forces (Ayantunde, Fernández-Rivera, and McCrabb, 2005), leaving herders to depend on locally available feed sources like pasture land, harvested fodder or crop residues—each being limited resources that can present major constraints to productivity (Kebreab, Smith, Tanner, and Osuji, 2005). Ensuring adequate feed to raise productive, healthy animals can be a challenge, especially during the dry season when herders are forced to trek up to 20 kms in search of good grazing conditions for their stock (KFSSG, 2013). Many pastoralists and livestock farmers therefore rely on forest resources as a coping strategy to overcome seasonal droughts that limit the availability of palatable grasses in natural pastures, either collecting the fodder themselves or taking cattle to graze directly within the reserves (Kiplagat, Mburu, and Mugendi, 2008; Wambugu, Franzel, Tuwei, and Karanja, 2001).

When it comes to gender relations, there are clear divisions regarding access to land and decision-making around both pasture and agricultural land use (Njuki and Sanginga, 2013b; Valdivia, 2001). Decision-making about livestock purchases, breeding and sale frequently rest firmly with men, while feeding and herding responsibilities overwhelmingly fall to women and children, both girls and boys (Verbeek, Kanis, Bett, and Kosgey, 2007). Kristjanson et al. (2010) suggest that securing the necessary inputs for healthy animals is often more difficult for women as they confront additional cultural and legal barriers to maintaining livestock grazing rights, while also paying increased opportunity costs to obtain feed and water, as locally available resources dwindle and degrade. In the past, common property grazing regimes, such as group ranches, have tended to exclude women from membership and decision-making (Mwangi, 2007; Ntiati, 2002), a trend that continues among contemporary forest user groups, as women tend to make up far less than half of community association members and rarely hold leadership roles (Coleman and Mwangi, 2013). Kenya Forest Service sources report that men are usually responsible for paying fees associated with fodder harvest permits and grazing licenses (pers. comm., Machakos Forest Service Administrator Benjamin Kineri, July 7, 2014), which is not surprising, as previous research suggests that women often lack the necessary economic resources to purchase animal feed and tend to have less access to these supplementary feed sources than their male counterparts (Miller, 2011).

As discussed in Box 8.3, the legalities of renting forestland for agricultural purposes are not firmly fixed, and are likely to change with time. According to the 2012 Forests Fees and Charges supplement, the annual rental fee is 500 KSh ($6 CAD) per acre, a sum substantially lower than what was reported for participation in the past shamba system (Witcome and Doward, 2009). Their results suggest a highly variable annual price that averaged closer to 1 000 KSh ($12 CAD) per acre, and included a required initial lump sum payment, but not the 'tips' (or bribes) necessary to secure high quality land of a desirable size. The results from existing pilot programs suggest that a new farmland renting system (also known as the 'Plantation Establishment and Livelihood Improvement Scheme') will be scaled-up eventually, but, as of 2015, no specific information has been forthcoming on the regulatory framework, implementation or enforcement that might accompany the programme.

Livestock grazing and fodder collection are both clearly defined in the Forests Act (2005) and the 2012 Forests Fees Supplement. Only cattle and sheep are permitted to enter the forest for on-site grazing at monthly costs of 100 KSh ($1.20 CAD) and 40 KSh ($0.50 CAD) respectively, per animal. Permits are only available during the dry season when tree seedlings are at less vulnerable growth stages and grazing is necessary to control the amount of undergrowth and dry brush in the forest, which in turn allows young trees to compete better for nutrients and sunlight, while also reducing risks of forest fire. Goats, camels and other livestock are prohibited due to their environmentally degrading feeding behavior, which can damage pastures and saplings and negatively affect ecosystem function. Collecting grass to feed livestock (and sometimes for sale at markets) costs 50 KSh ($0.60 CAD) per bag or head-load. Foresters suggest that this may be an unfairly high price point, as grazing cattle consume significantly greater quantities of fodder over the course of their month-long entry permits to the forest. As mentioned above, some cost adjustments may occur on a case-by-case basis through negotiation with certain community forest associations to incentivise farmers to take part in fodder harvesting.

Discussion

This examination of existing literature on use of natural resources and land for sub-sistence and income generation in Kenya highlighted gendered issues relating to the categories of harvesters, the benefits they receive, and the obstacles they may face. In breaking down formal policies by specific activity, we have been able to identify a number of disparities in the regulation of forest product collection, which likely exac-erbate inequality within traditional gendered divisions of labour. Biases emerge as we note the complex permitting procedures, substantial licensing fees and complete prohibitions that frequently apply to women's livelihood activities and household chores. Distinct contrasts exist—from the difficulty of gaining permission to harvest wild fruits and ambiguous rules for collecting drinking water, to the clearly defined fees and well-explained policies for burning charcoal—and these disparities will influence how different user groups profit and subsist from access to forest resources.

In addition, issues with existing policies on many female-dominated activities place undue burden on women and limit their access to naturally occurring and relatively abundant natural resources. Women generally take less than men and are charged more, and this inequality likely serves to undermine the resilience of household food provisioning to the detriment of men, women and children, as well as the environment. When regulations are nonexistent or difficult to interpret, forest users have been faced with threats of physical force, fines and other consequences as a result of arbitrary exercise of bureaucratic power, corruption and lack of predictability of state authorities (Shumsky, Hickey, Johns, Pelletier, and Galaty, 2014b). At the same time, private extractive industries and local elites can take advantage of their privileged access, greater resources and existing biases in enforcement to circumvent regulations and to unfairly access forest resources (Colchester et al., 2006; Rosenbaum, 2004). Easy-to-interpret rules and boundaries, such as the legal protections of timber harvesters, commercial entities and shamba renters (generally men), are necessary to maintain sustainable, equitable management of common pool resources like forests (Ostrom, 1990). In this sphere, women are important stewards of local natural resources, and their inclusion not only has environmental consequences, but also influences choices and livelihood options that can bolster household food security.

Informal rules and community regulations that fill the gap left by confusing (or nonexistent) formal regulations for forest produce harvested predominately by women can add to the problem, creating a patchwork of 'red tape' from local actors, international treaties and corrupt agents that varies from village to village. In this context, the pluralism of regulations from government, NGOs and community sources creates a difficult operating environment for small-scale harvesters. These ever-changing access rules often reinforce gender-based discrimination by institutionalizing prejudices that may already exist in customary institutions and informal regulatory structures (Ingram, 2014). It is also important to consider the consequences that arise from the lack of clear formal regulation for activities predominately undertaken by women, such as powerlessness, dependency on unpredictable actors, and lack of capacity to enact change within the system. Zulu and Richardson (2013) discussed these topics in detail in their examination of charcoal production rules in Senegal, where threats against property and personal safety, arbitrary law enforcement to serve individual interests, corruption and bribery combined to limit economic benefits for small-scale producers and exacerbated marginalization of vulnerable groups, especially resource-poor women.

'Double standards' in forest policies can take a number of forms, such as the gender-bias discussed in this chapter, the consistent advantages that wealthy elites and well-connected actors may gain over rural communities and local forest users, or other unfair advantages based on ethnic group, family origin or religious affiliation. Larson and Ribot (2007) explain how institutionalised biases against the poorest community members can push disenfranchised groups towards illegal extraction activities (which may also have adverse conservation consequences) and reinforce exclusion from access, benefit sharing and decision-making. They go on to suggest that formal forest policies play a large role in the asymmetric distribution of forest benefits, but that merely amending laws with more equity-focused language is not sufficient to overcome engrained attitudes, unequal implementation and unfair enforcement (Larson and Ribot, 2007). In agreement with these observations, we consider the similar case of gendered forest access and the need for comprehensive legislative revision to empower women, fight discrimination and support future equality, echoing this collection's gender-transformative approach (see also Njuki and Sanginga, 2013a).

Moving past 'gender blindness' by concentrating on asymmetries in power and politics, from the household level all the way up to international resource treaties, requires directed research and nuanced policy dialogue. Editing existing regulations to remove entrenched vulnerability caused by gender bias in harvest, use and access policies is not going to enough. Other scholars have addressed some of the mechanisms through which gender inequities can be mitigated, with suggestions ranging from legal reforms to ensure formal land rights to increasing the participation of, and accountability to, women in governing bodies. Such mechanisms are applicable at multiple scales, from local interventions to large-scale institutional reforms and significant evolution of policies (Nwoye, 2007).

However, placing a 'gendered face' on the issue of forest policy does not come easily to what has traditionally been a male-dominated, 'top down,' profit-driven capitalist domain in Kenya, especially in terms of policy development and the regulation of resource use—which ultimately disadvantages women users. Replacing policy structures with adaptive, resilience-*building* measures that support the entire population is a start, but empowering the community to continue this work remains the true objective in the long term. In this chapter, we have provided evidence of gender bias in formal forest policy in Kenya and considered the implications for household food security and resilience. We believe there is a need for a transformational shift towards recognizing and valuing the ways that subsistence collection of natural resources, forest-based livelihood activities, and equitable access to land foster resilience for men and women within the semi-arid farming systems of Kenya. Figure 8.2 presents an analytical framework that we think could help guide future discussions on natural resource management through a gendered lens.

For environmental regulations like forest laws and policies, we must remember the stated goal of these acts and enforcement is generally related to conservation and sustainable use, aimed at maintaining the resource for future generations as well as for profit seekers. In Kenya, where forest-based ecosystem services like biodiversity, climate stabilization and prevention of soil erosion can have a major impact on residents' livelihoods, appropriate regulations clearly provide a necessary framework to coordinate usage rights. Recent projects that emphasise adaptive learning—especially "unlearning" previously promoted conclusions about gender,

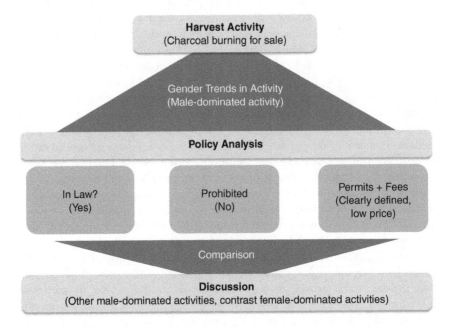

Figure 8.2 Gendered divisions in livelihoods. Women and men have traditionally exploited
different types of forest resources due to a variety of cultural, logistical and
regulatory reasons. As the Kenya Forest Service (KFS) defines and regulates most
(but not all) of these activities we can analyze their policies using a gendered
lens to better understand institutional biases and incongruent enforcement.

natural resource management and development interventions—have brought new
knowledge and perspective to a large audience of practitioners and researchers.
Ridgewell and colleagues' (2007a, 2007b) two volumes on natural resource management in the evolving and uncertain landscape of pastoral societies are examples
of the type of lesson-focused, open-minded research that will inform future policies and contribute to strengthening the kind of institutions where women are
truly empowered to participate in decision-making and creating formal regulations
that protect women's forest-based livelihood activities in ways that support multiple goals of household food security, income generation and nature conservation.

Endnote

1 Published most recently in 2012 as *The Forest (Fees and Charges) Rules.*

References

Adedayo, A., Oyun, M., and Kadeba, O. (2010). Access of rural women to forest resources
and its impact on rural household welfare in North Central Nigeria. *Forest Policy and
Economics,* 12, 439–450.

Agarwal, B. (2001). Participatory exclusions, community forestry, and gender: An analysis for South Asia and a conceptual framework. *World Development*, 29, 1623–1648.

Ajonina, P. U., Ajonina, G. N., Jin, E., Mekongo, F., Ayissi, I., and Usongo, L. (2005). Gender roles and economics of exploitation, processing and marketing of bivalves and impacts on forest resources in the Sanaga Delta region of Douala-Edea Wildlife Reserve, Cameroon. *International Journal of Sustainable Development and World Ecology*, 12, 161–172.

Arnold, J. E. M. and Perez, M. R. (2001). Can non-timber forest products match tropical forest conservation and development objectives? *Ecological Economics*, 39, 437–447.

Ayantunde, A., Fernández-Rivera, S., and McCrabb, G. (2005). *Coping with Feed Scarcity in Smallholder Livestock Systems in Developing Countries*. Nairobi: International Livestock Research Institute.

Behnke, R. and Muthami, D. (2011). *The Contribution of Livestock to the Kenyan Economy* (IGAD LPI Working Paper No. 03-11). Nairobi: IGAD Centre for Pastoral Areas and Livestock Development.

Bennett, E. M., Peterson, G. D., and Gordon, L. J. (2009). Understanding relationships among multiple ecosystem services. *Ecology Letters*, 12, 1394–1404.

Boyd, J. and Banzhaf, S. (2007). What are ecosystem services? The need for standardized environmental accounting units. *Ecological Economics*, 63, 616–626.

Brown, K. and Lapuyade, S. (2001). A livelihood from the forest: Gendered visions of social, economic and environmental change in southern Cameroon. *Journal of International Development*, 13, 1131–1149.

Burlingame, B. (2000). Wild nutrition. *Journal of Food Composition and Analysis*, 13, 99–100.

Byron, N. and Arnold, M. (1999). What futures for the people of the tropical forests? *World Development*, 27, 789–805.

Campbell, B. M. and Luckert, M. K. (2002). *Uncovering the Hidden Harvest: Valuation Methods for Woodland and Forest Resources*. London: Earthscan.

Castro, A. P. and Nielsen, E. (2003). *Natural Resource Conflict Management Case Studies: An Analysis of Power, Participation and Protected Areas*. Rome: Food and Agriculture Organization of the United Nations.

Cavendish, W. (2000). Empirical regularities in the poverty-environment relationship of rural households: Evidence from Zimbabwe. *World Development*, 28, 1979–2003.

Colchester, M., Boscolo, M., Contreras-Hermosilla, A., Del Gatto, F., Dempsey, J., Lescuyer, G. et al. (2006). *Justice in the Forest: Rural Livelihoods and Forest Law Enforcement*. Bogor: Center for International Forestry Research.

Coleman, E. A. and Mwangi, E. (2013). Women's participation in forest management: A cross-country analysis. *Global Environmental Change*, 23, 193–205.

Coulibaly-Lingani, P., Tigabu, M., Savadogo, P., Oden, P. C., and Ouadba, J. M. (2009). Determinants of access to forest products in southern Burkina Faso. *Forest Policy and Economics*, 11, 516–524.

Crafter, S., Awimbo, J., and Broekhoven, A. (1997). *Non-Timber Forest Products: Value, Use and Management Issues in Africa, Including Examples from Latin America* (Proceedings of a workshop held in Naro Moru, Kenya, 8–13 May 1994). Nairobi: International Union for Conservation of Nature.

Delang, C. O. (2006). Not just minor forest products: The economic rationale for the consumption of wild food plants by subsistence farmers. *Ecological Economics*, 59, 64–73.

Dovie, D. B., Witkowski, E. T., and Shackleton, C. M. (2005). Monetary valuation of livelihoods for understanding the composition and complexity of rural households. *Agriculture and Human Values*, 22, 87–103.

Ellis, F. and Allison, E. (2004). *Livelihood Diversification and Natural Resource Access* (Livelihood Support Programme Working Paper No. 9). Rome: Food and Agriculture Organization of the United Nations.

Emerton, L. (1996). *Valuing the Subsistence Use of Forest Products in Oldonyo Orok Forest, Kenya* (Rural Development Forestry Network Paper 19e). London: Overseas Development Institute.

Eriksen, S. H., Brown, K., and Kelly, P. M. (2005). The dynamics of vulnerability: Locating coping strategies in Kenya and Tanzania. *Geographical Journal*, 171, 287–305.

FAO. (2005). *Global Forest Resources Assessment*. Rome: Food and Agriculture Organization of the United Nations.

FAO. (2010). *Global Forest Resources Assessment: Key Findings*. Rome: Food and Agriculture Organization of the United Nations.

FAO. (2012). *State of the World's Forests*. Rome: Food and Agriculture Organization of the United Nations.

FAO. (2014). *State of the World's Forests*. Rome: Food and Agriculture Organization of the United Nations.

Feka, N. Z., Manzano, M. G., and Dahdouh-Guebas, F. (2011). The effects of different gender harvesting practices on mangrove ecology and conservation in Cameroon. *International Journal of Biodiversity Science, Ecosystem Services and Management*, 7, 108–121.

Fisher, M. (2004). Household welfare and forest dependence in Southern Malawi. *Environment and Development Economics*, 9, 135–154.

Folke, C. (2004). Traditional knowledge in social-ecological systems [editorial]. *Ecology and Society*, 9, 7.

Haro, G. O., Doyo, G. J., and McPeak, J. G. (2005). Linkages between community, environmental, and conflict management: Experiences from Northern Kenya. *World Development*, 33, 285–299.

Harris, F. M. A. and Mohammed, S. (2003). Relying on nature: Wild foods in Northern Nigeria. *AMBIO: A Journal of the Human Environment*, 32, 24–29.

Hasalkar, S. and Jadhav, V. (2004). Role of women in the use of non-timber. Forest produce: A review. *Journal of Social Science*, 8, 203–206.

Helander, H. and Larsson, L. (2014). *Emissions and Energy Use Efficiency of Household Biochar Production during Cooking in Kenya* (Bachelor's thesis). Uppsala University, Sweden.

Ingram, V. (2014). *Win-Wins in Forest Product Value Chains? How Governance Impacts the Sustainability of Livelihoods Based on Non-Timber Forest Products from Cameroon*. Leiden: African Studies Centre.

Iwu, M. M. (2014). *Handbook of African Medicinal Plants, Second Edition*. New York, NY: CRC press.

Jackson, C. (1993). Doing what comes naturally? Women and environment in development. *World Development*, 21, 1947–1963.

Kabubo-Mariara, J. (2013). Forest-poverty nexus: Exploring the contribution of forests to rural livelihoods in Kenya. *Natural Resources Forum*, 37: 177–188.

Kalame, F. B. (2011). *Forest Governance and Climate Change Adaptation: Case Studies of Four African Countries* (Tropical Forestry Reports 39). Helsinki: Viikki Tropical Research Institute.

Kebreab, E., Smith, T., Tanner, J., and Osuji, P. (2005). Review of undernutrition in smallholder ruminant production systems in the tropics. Nairobi: International Livestock Research Institute.

KFS. (2009). *2009/10–2013/14 Strategic Plan: Trees for Better Lives*. Nairobi: Kenya Forest Service.

KFS. (2012). *The Forests (Fees and Charges) Rules* (Kenya Gazette Supplement No. 132; Legislative Supplement No. 38; Legal Notice No. 104). Nairobi: Kenya Forest Service.

KFSSG. (2013). *The 2012–2013 Short Rain Season Food Assessment Report*. Nairobi: Kenya Food Security Steering Group.

Kiplagat, A., Mburu, J., and Mugendi, D. (2008). *Consumption of Non Timber Forest Products (NTFPs) in Kakamega Forest, Western Kenya: Accessibility, Role and Value to Resident Rural Households* (IASC Biennial International Conference 14–19 July 2008). Cheltenham: University of Gloucestershire.

Kokwaro, J. O. (2009). *Medicinal Plants of East Africa*. Nairobi: University of Nairobi Press.

Kristjanson, P., Waters-Bayer, A., Johnson, N., Tipilda, A., Baltenweck, I. et al. (2010). *Livestock and Women's Livelihoods: A Review of the Recent Evidence* (ILRI Discussion Paper No. 20). Nairobi: International Livestock Research Institute.

Kusters, K., Achdiawan, R., Belcher, B., and Ruiz Pérez, M. (2006). Balancing development and conservation? An assessment of livelihood and environmental outcomes of nontimber forest product trade in Asia, Africa, and Latin America. *Ecology and Society*, 11, 20.

Larson, A. M. and Ribot, J. C. (2007). The poverty of forestry policy: Double standards on an uneven playing field. *Sustainability Science*, 2, 189–204.

Leach, M. (2007). Earth mother myths and other ecofeminist fables: How a strategic notion rose and fell. *Development and Change*, 38, 67–85.

Maathai, W. (2004). *The Green Belt Movement: Sharing the Approach and the Experience.* New York, NY: Lantern Books.

Mahiri, I. and Howorth, C. (2001). Twenty years of resolving the irresolvable: Approaches to the fuelwood problem in Kenya. *Land Degradation and Development*, 12, 205–215.

Mai, Y., Mwangi, E., and Wan, M. (2011). Gender analysis in forestry research: Looking back and thinking ahead. *International Forestry Review*, 13, 245–258.

Mamo, G., Sjaastad, E., and Vedeld, P. (2007). Economic dependence on forest resources: A case from Dendi District, Ethiopia. *Forest Policy and Economics*, 9, 916–927.

Matiru, V. (2000). Forest cover and forest reserves in Kenya: Policy and practice (Forest and Social Perspectives in Conservation Working Paper). Nairobi: IUCN Regional Office for Eastern Africa.

Maundu, P. M, Ngugi, G. W., and Kabuye, C. H. S. (1999). *Traditional Food Plants of Kenya.* Nairobi: National Museum of Kenya.

Mbuvi, D. and Boon, E. (2009). The livelihood potential of non-wood forest products: The case of Mbooni Division in Makueni District, Kenya. *Environment, Development and Sustainability*, 11, 989–1004.

Meinzen-Dick, R. S., Brown, L. R., Feldstein, H. S., and Quisumbing, A. R. (1997). Gender, property rights, and natural resources. *World Development*, 25, 1303–1315.

Miller, B. A. (2011). *The Gender and Social Dimensions to Livestock Keeping in Africa: Implications for Animal Health Interventions.* Edinburgh: GALVmed.

Monbiot, G. (1994). The tragedy of enclosure. *Scientific American*, 270159.

Mukonyi, K. and Lwande, W. (2007). Bioprospecting for wealth and biodiversity conservation: Case of Mondia Whytei commercialization around Kakamega forest ecosystem in Kenya, *Discovery and Innovation*, 19, Special edition 4, 389–397.

Muli, E., Munguti, A., and Raina, S. (2007). Quality of honey harvested and processed using traditional methods in rural areas of Kenya. *Acta Veterinaria Brno*, 76, 315–320.

Murton, J. (1999). Population growth and poverty in Machakos District, Kenya. *Geographical Journal*, 165, 37–46.

Muyanga, M. and Jayne, T. S. (2006). *Agricultural Extension in Kenya: Practice and Policy Lessons* (Working Paper No. 26). Nairobi: Tegemeo Institute of Agricultural Policy and Development, Egerton University.

Mwangi, E. (2007). The puzzle of group ranch subdivision in Kenya's Maasailand. *Development and Change*, 38, 889–910.

Mwenda, A. and Kibutu, T. (2012). Implications of the new constitution on environmental management in Kenya. *Law, Environment and Development Journal*, 8, 76–88.

Neumann, R. P. and Hirsch, E. (2000). *Commercialisation of Non-Timber Forest Products: Review and Analysis of Research.* Bogor: Center for International Forestry Research.

Njoroge, G., Bussmann, R., Gemmill, B., Newton, L., and Ngumi, V. (2004). Utilization of weed species as source of traditional medicines in central Kenya. *Lyonia* 7, 71–87.

Njoroge, K. (2011, March 14). Kenya to reintroduce shamba system. *Standard Digital*. Retrieved from https://www.standardmedia.co.ke

Njuki, J. and Sanginga, P. C. (Eds.). (2013a). *Women, Livestock Ownership and Markets: Bridging the Gender Gap in Eastern and Southern Africa*. Abingdon: Routledge.

Njuki, J. and Sanginga, P. (2013b). Gender and livestock: Key issues, challenges and opportunities. In Njuki, J., and Sanginga, P. C. (Eds.), *Women, Livestock Ownership and Markets: Bridging the Gender Gap in Eastern and Southern Africa* (pp. 1–8). Abingdon: Routledge.

Ntiati, P. (2002). *Group Ranches Subdivision Study in Loitokitok Division of Kajiado District, Kenya* (Land Use Change Impacts and Dynamics Project Working Paper 19). Nairobi: International Livestock Research Institute.

Nwoye, M. (2007). Gender responsive entrepreneurial economy of Nigeria: Enabling women in a disabling environment. *Journal of International Women's Studies*, 9, 167–175.

Nzengya, D. M. and Aggarwal, R. (2013). Water accessibility and women's participation along the rural-urban gradient: A study in Lake Victoria Region, Kenya. *Journal of Geography and Regional Planning*, 6, 263–273.

Opiyo, F., Wasonga, O., Schilling, J., and Mureithi, S. (2012). Resource-based conflicts in drought-prone Northwestern Kenya: The drivers and mitigation mechanisms. *Wudpecker Journal of Agricultural Research*, 1, 442–453.

Ostrom, E. (1990). *Governing the Commons: The Evolution of Institutions for Collective Action*. Cambridge: Cambridge University Press.

Ouma, E. A., Obare, G. A., and Staal, S. J. (2003). Cattle as assets: Assessment of non-market benefits from cattle in smallholder Kenyan crop-livestock systems. *Proceedings of the 25th International Conference of Agricultural Economists (IAAE)*, 16–22.

Pandolfelli, L., Dohrn, S., and Meinzen-Dick, R. S. (2007). *Gender and Collective Action: Policy Implications from Recent Research* (Policy Brief No. 5). Washington, DC: CGIAR Systemwide Program on Collective Action and Property Rights.

PROCASUR. (2012). An overview of livestock sub-sector in Kenya. Retrieved from http://www.cop-ppld.net/cop_knowledge_base/detail/?dyna_fef%5Buid%5D=3437

Quisumbing, A. R. (2003). *Household Decisions, Gender, and Development: A Synthesis of Recent Research*. Washington, DC: International Food Policy Research Institute.

Republic of Kenya. (2000). *Environmental Management and Coordination Act, EMCA* (Act No. 8 of 1999, Kenya Gazette Supplement No. 3, Acts No. 1).

Republic of Kenya. (2005). *The Kenya Forests Act, 2005*. Ministry of Environment and Natural Resources, Kenya.

Republic of Kenya. (2010). *Constitution of Kenya*.

Ridgewell, A., Mamo, G., and Flintan, F. (Eds.). (2007a). *Gender and Pastoralism Vol. 1: Rangeland and Resource Management in Ethiopia*. Addis Ababa: SOS Sahel Ethiopia.

Ridgewell, A. and Flintan, F. (Eds.). (2007b). *Gender and Pastoralism Vol. 2: Livelihoods and Income Development in Ethiopia*. Addis Ababa: SOS Sahel Ethiopia.

Robinson, E. J. Z. and Lokina, R. B. (2011). A spatial, temporal analysis of the impact of access restrictions on forest landscapes and household welfare in Tanzania. *Forest Policy and Economics*, 13, 79–85.

Rocheleau, D. and Edmunds, D. (1997). Women, men and trees: Gender, power and property in forest and agrarian landscapes. *World Development*, 25, 1351–1371.

Ros-Tonen, M. A. F. and Wiersum, K. F. (2005). The scope for improving rural livelihoods through non-timber forest products: An evolving research agenda. *Forests, Trees and Livelihoods*, 15, 129–148.

Rosenbaum, K. L. (2004). Illegal actions and the forest sector. *Journal of Sustainable Forestry*, 19, 263–291.

Scoones, I., Melnyk, M., and Pretty, J. N. (1992). The hidden harvest—Wild foods and agricultural systems: A literature review and annotated bibliography. London: International Institute for Environment and Development, Sustainable Agriculture Programme.

Shackleton, C. and Shackleton, S. (2004). The importance of non-timber forest products in rural livelihood security and as safety nets: a review of evidence from South Africa. *South African Journal of Science*, 100, 658–654.

Shackleton, S., Paumgarten, F., Kassa, H., Husselman, M., and Zida, M. (2011). Opportunities for enhancing poor women's socioeconomic empowerment in the value chains of three African non-timber forest products (NTFPs). *International Forestry Review*, 13, 136–151.

Shumsky, S. A., Hickey, G. M., Pelletier, B., and Johns, T. (2014a). Understanding the contribution of wild edible plants to rural social-ecological resilience in semi-arid Kenya. *Ecology and Society*, 19, 34.

Shumsky, S., Hickey, G. M., Johns, T., Pelletier, B., and Galaty, J. (2014b). Institutional factors affecting wild edible plant (WEP) harvest and consumption in semi-arid Kenya. *Land Use Policy*, 38, 48–69.

Sihanya, B. (2011). Constitutional implementation in Kenya, 2010–2015: Challenges and prospects. (FES Kenya Occasional Paper No. 5). Nairobi: Friedrich-Ebert-Stiftung.

Sunderland, T., Achdiawan, R., Angelsen, A., Babigumira, R., Ickowitz, A. et al. (2014). Challenging perceptions about men, women, and forest product use: A global comparative study. *World Development*, 64, S56–S66.

Termote, C. (2011). Eating from the wild: Turumbu, Mbole and Bali traditional knowledge on non-cultivated edible plants, District Tshopo, DR Congo. *Genetic Resources and Crop Evolution*, 58, 585.

Turner, T. E. and Brownhill, L. (2006). Ecofeminist action to stop climate change. Paper presented at the International Society for Ecological Economics Ninth Biennial Conference, Delhi, India, Dec. 15–18.

United Nations. (2005). *Millennium Ecosystem Assessment: Ecosystems and Human Well-Being—Our Human Planet: Summary for Decision Makers.* Washington, DC: Island Press.

Valdivia, C. (2001). Gender, livestock assets, resource management, and food security: Lessons from the SR-CRSP. *Agriculture and Human Values*, 18, 27–39.

Verbeek, E., Kanis, E., Bett, R., and Kosgey, I. (2007). Socio-economic factors influencing small ruminant breeding in Kenya. *Livestock Research for Rural Development*, 19, n.p.

Veuthey, S. and Gerber, J. F. (2010). Logging conflicts in Southern Cameroon: A feminist ecological economics perspective. *Ecological Economics*, 70, 170–177.

Vira, B., Wildburger, C., and Mansourian, S. (Eds.). (2015). *Forests, Trees and Landscapes for Food Security and Nutrition: A Global Assessment Report* (IUFRO World Series Vol. 33). Vienna: International Union of Forest Research Organizations.

Wambugu, C., Franzel, S., Tuwei, P., and Karanja, G. (2001). Scaling up the use of fodder shrubs in central Kenya. *Development in Practice*, 11, 487–494.

Were, E. A., Swallow, B. M., and Roy, J. (2006). Water, women, and local social organization in the western Kenya highlands. (Working Paper No. 51). Washington, DC: CGIAR Systemwide Program on Collective Action and Property Rights.

Witcomb, M. and Dorward, P. (2009). An assessment of the benefits and limitations of the shamba agroforestry system in Kenya and of management and policy requirements for its successful and sustainable reintroduction. *Agroforestry Systems*, 75, 261–274.

Yatich, T., Awiti, A., Nyukuri, E., Mutua, J., Kyalo, A., Tanui, J., and Catacutan, D. (2007). Policy and institutional context for NRM in Kenya: Challenges and opportunities for Landcare (ICRAF Working Paper No. 43). Nairobi: World Agroforestry Centre.

Zulu, L. C. (2010). The forbidden fuel: Charcoal, urban woodfuel demand and supply dynamics, community forest management and woodfuel policy in Malawi. *Energy Policy*, 38, 3717–3730.

9 Accountability and citizen participation in devolved agricultural policy-making

Insights from Makueni County, Kenya

Leigh Brownhill, Tony Moturi, and Gordon M. Hickey

This chapter considers changes in food security–related policy-making in Kenya after the national adoption of a devolved governance framework in 2010. Participation, as this chapter tries to highlight, is as much an important part of effective research as it is central to the devolution of governance. Many scholars have identified participatory approaches in research as a positive step towards more responsive and effective science, policy and agricultural development (Chambers and Jiggins, 1987; Ashby and Sterling, 1995; Pretty, 1995; White, 1996; Sanginga, Chitsike, Njuki, Kaaria, and Kanzikwera, 2007; Cornwall, 2008; Neef and Neubert, 2011). Yet such practices are far from being widely implemented (Hall and Yoganand, 2004). This is especially true when it comes to women's participation in development, extension, research and policy-making. The gap between the rhetoric of farmer participation and its implementation in research and policy processes suggests the presence of important institutional and political constraints, including entrenched gender inequalities and insufficient information flows (Brownhill and Hickey, 2012; Eidt, Hickey, and Curtis, 2012).

With political goodwill, and based on available natural and social resources, policy can support and mobilise local livelihood strategies, which, in turn, can build local socio-ecological resilience. Increasing poor farmers' access to resources, and the livelihood strategies that agricultural policies can support, is perhaps the shortest and most direct route towards building food security and farming system resilience. Farmer participation, insofar as it improves access to information, resources, research and policy-making, takes on a special significance as a first-priority goal which can facilitate all other aspects of enhanced food security and agricultural development more broadly.

In what follows we assess the scope and strength of the existing, emerging and planned mechanisms through which public participation in local governance is enabled in Kenya. Our work draws on research into the perceptions, experience and aspirations of farmers and other agriculture stakeholders in the semi-arid county of Makueni, as a window into farmers' engagement in local agricultural policy processes, both before and since the promulgation of devolution. We then consider the Makueni case in light of the literature on decentralisation and devolution in Africa and more broadly. We aim to demonstrate the considerable capacities and aspirations of the local people whom devolution is meant to benefit. Their aspirations indicate a readiness for civic engagement; and their diverse capacities together provide measures of local interests, priorities and concerns. We conclude with suggestions for strengthening and catalysing 'participation' into more effective and locally relevant policies to help build socio-ecological resilience at the county level.

Devolution defined

Devolution, in general, is a form of democratic decentralisation. In its application internationally, devolution has been described as having "two principle dimensions: (a) top-down measures aimed at transferring responsibilities—political, administrative and/or fiscal—to lower levels of government and (b) the gradual opening of spaces for participation from below" (Larson and Soto, 2008, p. 216). How these changes are enacted significantly shapes not only how governance is done, but also the nature of the outcomes (results and consequences) of the decisions made. Larson and Soto distinguish between 'technocratic' and 'transformational' forms of decentralisation. "*Technocratic* decentralization is primarily aimed at increasing efficiency and building transparency and institutional stability as essential governance conditions for the promotion of private investment" (Larson and Soto, 2008, p. 230). In contrast, decentralisation that is *transformational* "challeng[es] existing power relations rather than prioritizing efficient service delivery. ...transform[s] and democratize[s] the political process itself" and structurally separates "the accumulation of political and economic power" (Hickey and Mohan, 2005, cited in Larson and Soto, 2008, p. 231). But as with participation, there is often a gap between the rhetoric and the reality of devolution. In a review of decentralisation processes in Africa, Ribot identified the common contradiction between the intention for devolution and the maintenance of a hierarchical status quo: "Because decentralizations that democratize and transfer powers threaten many actors, few have been fully implemented" (2002, p. v). These actors include many local political and economic elites, as well as international actors who have had considerable influence over policy-making in the past.

Our line of inquiry and review of literature on devolved agricultural policy-making shed light on two interdependent elements of 'successful' devolution: public participation and accountability. Both can be described with reference to a spectrum of attributes (e.g., deep, shallow, upward, horizontal, downward), depending on the intended outcome and the practices employed. And while both are needed for implementing devolved governance, we also trace a relationship between them that suggests that accountability is reliant upon public participation.

We turn now to a very brief review of devolution's evolution in Kenya, after which we explore the diverse forms that participation and accountability can take. We then present a case study from Makueni and draw insights from it that may be useful in ongoing efforts to structure and facilitate public participation.

Devolution in Kenya

Top-down decision-making structures in Kenya are rooted in Britain's geographically uneven imposition of colonial rule in the early 20th century. Colonial-era infrastructural development, including roads, railway spurs, markets, hospitals, schools and public offices, was concentrated along a 150 km strip straddling the cross-country railway the British had built between 1896 and 1901. "What had emerged by the year 2000 was an intense concentration of political and administrative power, wealth and population along the railway line. Except Garissa, all provincial headquarters were either on the railway line or within 40 kilometres of it. So were 80 per cent of the population and an even larger proportion of the nation's wealth" (Wachira, 2013).

Although British rule ended in Kenya more than 50 years ago, much of the infrastructural and administrative structure remains. The colonial administrative network of interlinked villages, sub-locations, locations, districts, divisions and provinces long ago laid down structures of 'indirect rule.' These brought government 'close to the people,' but in top-down, hierarchical and often punitive relationships. Thus as state and citizens democratise governance institutions through devolution, it is the *content of the relations of ruling* which are changing rather than simply the installation of new government offices or administrative levels.

Efforts were made to make government more responsive to, and representative of, citizens after the declaration of independence in 1963. However, the Sessional Paper No. 10 of 1963 on the application of African socialism in Kenya rationalised unequal development by prioritising social and economic investments in development of high-potential areas, to the exclusion of other regions, especially arid and semi-arid lands such as those in Makueni. By the 1970s, decentralisation had become a "cornerstone" of Kenya's development planning. "The goal was to 'coordinate and stimulate development at the local level by involving in the planning process not only government officials but also the people *through their representatives.'* Kenya established a system of district development committees in 1974 through which technical assistance was provided to local planning organizations." Despite the installation of these structures intended to decentralise planning, by 1980, "control over development planning and administration remain[ed] highly centralized" (Rondinelli, 1980, p. 134).

This centralisation of political power was in part a legacy of the tight integration of administrative and economic power begun under British rule and continued after independence. "Local leaders and large landowners often form alliances with ministry officials and members of parliament to protect current patterns of decision-making or to resist changes proposed by district development committees that are adverse to their interests" (Rondinelli, 1980, p. 140). Such arrangements between economic and political elites often kept the development committees from better representing people at the grassroots. At the same time, successive amendments to the constitution in the 1970s and 1980s concentrated decision-making power in the Executive. This concentration of power was exacerbated by the top-down policy directives of the World Bank and International Monetary Fund in their structural adjustment programmes beginning in 1980, which introduced very unpopular policy measures as conditions for loans and aid (Stein and Nissanke, 1999).

Public perception of an increasing concentration and abuse of power in Kenya motivated a twenty-year popular democracy movement, which, in the 1990s, vociferously demanded wide-ranging democratic reforms, starting with the repeal of Section 2(a), an amendment to the constitution that had banned opposition political parties. A strong government crackdown on pro-democracy activists became a rallying point in national and international circles, culminating in the "1991 Paris Consultative Group's decision to withhold $1bn aid pending political reforms" (Ajulu, 1998, p. 275). The international financial institutions and international development organisations pursued campaigns of privatisation at the same time as they promoted liberalisation, democratisation and good governance (Ribot, 2002, p. iv). In these campaigns, the call for political democratisation echoed from above what the growing mobilisations of Kenyan citizens had long been building alliances to demand from below (grassroots) and from "the middle" (civil society organisations) (Mutunga, 1999). With the support of an increasingly alarmed international diplomatic community, the government repealed Section 2(a) in 1991.

The democracy movement that saw Section 2(a) repealed did not stop with this small but significant victory. The legalisation of opposition political parties emboldened grassroots activists and fostered popular organising for more wide-ranging constitutional change and social transformation (Brownhill, 2009). Democratic devolution was at the heart of popular demands. By 1997, President Daniel arap Moi gave in to popular pressure and instituted a formal process for the drafting of a new constitution. Kenyans came out in large numbers to vote in the referendum on the new constitution, which was promulgated in August 2010.

The main objectives of the devolution of government in Kenya are stated in Article 174 of the Constitution of Kenya (2010), and include diversity, national unity, democracy, transparency, accountability, promotion of social and economic development, promotion of people's participation in decisions that affect them and promotion of the equitable distribution of resources. Here the constitution provides signposts towards the mobilisation of effective citizen participation in devolved governance.

The constitution's Fourth Schedule (Article 185(2), 186(1) and 187(2)) describes the distribution of functions between the national government and the county governments. With regard to agriculture, the national government retains control over policy formulation and the setting of uniform standards and norms; while the function of agricultural service provision and policy implementation falls to county governments. These services include crop and animal husbandry, livestock sale yards, county abattoirs, plant and animal disease control, and fisheries. Although agricultural policy is made at the national level, localising or domesticating policies at county levels is envisioned. The Transition Authority, which is legally mandated to assign functions to either level of government, has broken down or unbundled these services into actual deliverable actions and assigned costs to them. It is on the basis of this unbundling and costing that revenue is allocated to counties to deliver agricultural services. The constitutional principle underpinning this unbundling and costing is that 'resources must follow functions.' This brings us to a further consideration of key elements of devolved decision-making that promise to open space for direct, democratic citizen participation.

Participation

Devolution implies a move from representative to a more direct form of demo-cratic governance. But how do we measure participation? Drawing on participatory research methods, Cornwall reviewed a "simple axis" devised to gauge "forms of participation according to depth and breadth."

> *A 'deep' participatory process engages participants in all stages of a given activity, from identification to decision-making. Such a process can remain 'narrow,' however, if it only involves a handful of people, or particular interest groups. Equally, a 'wide' range of people might be involved, but if they are only informed or consulted their participation would remain 'shallow'. This usefully highlights the intersections between inclusion/exclusion and degrees of involvement* (Cornwall, 2008, p. 276).

While devolving states can employ this axis to measure and try to attain deeper participation, there are also distinctions among states that should be taken into account when considering how to facilitate citizen engagement. In a review of what hampers and what enables the participation of the rural poor, Sumner et al. (2008) distinguished the processes characterising 'Northern' and 'Southern' states. They identified a degree of "complexity" in policy processes in the global south as "a product of the interaction of both Northern and Southern contexts. Development policy involves donors—that is, Northern contexts—as well as actors in developing countries—that is, Southern contexts. On the one hand 'ownership' via budget support is shifting the emphasis to Southern contexts. On the other hand donors still exert considerable visible and invisible influence" (Sumner, Acosta, Kapur, Bahadur, and Bobde, 2008, pp. 5–6). Similarly, Shayo et al., in addressing decision-making and priority-setting for devolved health policy in Tanzania, show that donor and ministry officials had undue influence in the setting of priorities, which reduced the effectiveness of public participation. In this case, a top-down approach led to less effective and less inclusive policies: "…the system remained heavily reliant on external funders who tend to guard their own globally gener-ated priority agendas in a manner that disregards local experiences and priorities" (Shayo, Mboera, and Blystad, 2013, p. 8). Kenyans, then, might need to carefully consider how citizens' priorities can begin to outweigh the agendas of the histori-cally more powerful external funders. Commitment to deeper participation by a more diverse array of citizens could be one step towards this rebalancing of power.

Devolution can be especially empowering for women. In the South African and Ugandan settings, devolution has facilitated the political participation of women especially at lower, local levels: "As in Jinja [Uganda], in East London [South Africa] women are the most active in community organisations and so participate more actively in development forums and, combined with women councillors, their par-ticipation created a critical mass of women in politics in East London.…women maintain a clear focus on development issues, such as housing, health and educa-tion. They attend meetings regularly [and] are very active in organising" (Dauda, 2006, p. 299). Bardhan identified the power of women's political engagement in

India, even if confined to lower levels of governance: "the women leaders of village councils invest more in infrastructure that is directly relevant to the needs of rural women, like drinking water, fuel and roads, and that village women are more likely to participate in the policy-making process if the leader of their village council is a woman" (Bardhan, 2002, p. 198). Thus women's concentration at lower levels of governance might be taken as a foundation on which to build and enhance public engagement in devolved governance, especially by this often-marginalised group.

Accountability

A wide literature on decentralisation and devolution policies identifies accountability and public participation as two seminal practices through which the direction of devolution is decided. This is recognised explicitly in the Kenyan government's own report on devolution (Government of Kenya [GOK], 2011). As participation can take a range of forms, so, too, does accountability. In their study of citizen participation and accountability in Kenya and Uganda, Devas and Grant (2003) noted three forms of accountability: "upward accountability" (e.g., local to central government); "horizontal accountability" (e.g., local government officials to elected representatives); and "downward accountability" (e.g., elected representatives to local citizens). Of the three, 'downward accountability,' they found, was least often achieved, reflecting insufficient participation and resulting in difficulties in eradicating corruption.

Southern governments' officials, both elected and appointed, are often pressured to be upwardly accountable for loans and donor funding. Such directives are followed at least in part because they come from those who hold the purse-strings, so to speak, and can pull them closed if unsatisfied. Devas and Grant identify a risk that "the emphasis on upward accountability impedes the development of downward accountability, as local officials and elected representatives devote their attention to meeting external performance conditions and can hide behind central government funding requirements as an excuse for failing to deliver to local citizens" (Devas and Grant, 2003, p. 315). A similar tension arises with foreign investors, who are legally bound to account to their shareholders, rather than to the citizens of the countries in which they invest. Yet in Kenya, as in many countries undergoing devolution, local officials at county levels are empowered to seek independent sources of finance, and many are looking to investors who, in the end, are not bound by democratic mores prevailing in Kenya. "Investors are looking intently at the prospects for access to vast mineral, oil and gas deposits. Now they will have to negotiate with the county governments" (Kantai, 2013). Such eventualities bring with them serious questions about who will decide about such projects, and how all levels of accountability will be ensured, especially downwardly. The distinction between accountability to shareholders and to citizens was recognised by Senator Billow Kerrow of Mandera County, Kenya, in a March 2015 review of progress in implementing devolved governance two years after the 2013 election: "Public service is not the corporate world; there is accountability to the court of public opinion" (Kerrow, 2015). So while public servants are bound by accountability to the public, the private investors they are inviting into their counties are not, and conflicts of interest are inevitable.

Crawford's 2009 study of decentralisation in Ghana found that "the key to enhanced local democracy is the strengthening of downward accountability mechanisms, yet the case study also indicates that such reforms are not easily achieved." He attributed this shortcoming to an "unwillingness" of central governments to relinquish control to devolved institutions (Crawford, 2009, pp. 57–58). Considered alongside Devas and Grant's findings, above, it appears that accountability to the public can be strengthened with focus, from below, on greater organisation for participation, and from above, on action by external funders and government officials, both local and central, to democratise political and economic relations (Cornwall, 2008, p. 278). For government officials, formal processes of accountability and zero tolerance for corruption offer clear ways forward. But for external funders and foreign investors, accountability to the Kenyan public is a less straightforward matter, for these parties hold a more tangential relationship to the citizens in countries in which they allocate funds or invest; they 'deal' directly with governing officials authorised to take on debt or welcome investors. The responsibility, then, for donor and investor accountability to the Kenyan public may fall onto the shoulders of local citizens, who may realise their rights to engage in decision-making and hold both governing officials and investors accountable to the public.

Ultimately, accountability results from democratic relationships, fostered through citizen participation and facilitated by the free flow of information. In what follows we consider some of the ways that devolution, incorporating principles of participation and accountability, might be realised in the agriculture sector, drawing on the experiences and aspirations of smallholder farmers in Kenya.

Case study: Devolved agricultural policy-making in Makueni

To understand existing opportunities for, and barriers to, devolved agricultural policy-making in semi-arid Eastern Kenya, we conducted some 45 interviews and six focus group discussions (FGD) in Makueni County, specifically in Wote, Kaiti and Mbooni. An iterative and exploratory approach was employed to collect data from key informants following a snowball sampling strategy between August 2010 and September 2014. At an early stage of the research, we examined the sharing and dissemination of information through 'interview triads' with researchers, policy-makers and farmers (Brownhill and Hickey, 2012). We continued, from there, to consider the experiences and aspirations of farmers within a larger matrix of farm system stakeholders involved in policy-making processes. Purposive selection of an initial set of respondents led to the identification of further participants, including farmer group members, agricultural researchers, county government officials, cooperative society representatives, community-based organisations' leaders, religious leaders, civic educators, youth leaders and community activists. Interviews and focus groups explored participants' views on and experiences with agricultural policy-making, and sought their narrations of "what is," "what ought to be" and "how to get from here to there."

The small scale and subsistence farmers we interviewed in Makueni County expressed a strong interest in engaging more fully in setting priorities such that the government policies and services could better meet local needs and interests. Farmers generally expressed a feeling that their experiences and interests were, for the most part, left out of local policy-making processes, noting in our discussions that "the government does not consult the rural people about their problems," and that "the farmers are not involved in any way in agricultural policy-making" (Wote, FGD, 8 March 2012). Most reported that they felt like recipients of policy prescriptions rather than active agents of agricultural development. Yet, access to information was reported as lacking: "through government or any other organization dealing with agriculture, the approach of passing the information to farmers is very poor and ineffective." Respondents indicated that policy prescriptions were likely to find more acceptance and compliance among farmers in the future if the farmers were engaged in shaping the direction of change and thus "owning" and seeing the benefits (or lack of same) of the policies and programmes that result.

An extension officer in Wote told us that she expected, with devolution, that "everything will be done just around the community, and the leaders will interact with us directly and they will follow the wishes of the localities" (Interview, 23 June 2011). Mandera Senator Bill Kerrow echoed this popular expectation of devolution when he emphasised that "devolution is also about democracy at grassroots that empowers residents to express themselves to determine their future" (Kerrow, 2015). These are expected arrangements. In practice, one Kenyan senator critiqued the lack of clear guidance on participation. While the Transition Authority and the Ministry of Devolution and Planning have produced participation guidelines, they are not anchored in law and fully recognised by different government parties, or the public. This means that in practice, "county governments go around markets collecting public views on its development proposals," a practice which "is both inefficient and ineffective" (Nyong'o, 2015).

County agriculture department officials involved in implementing agricultural policies and providing services are among the most visible and accessible 'policy-makers' who interact with farmers on a regular basis. Respondents emphasised that for decades, extension services have been under-funded, and thus fewer and fewer farmers ever interact with experts (Interview, 23 June 2011). A 'train and visit' programme formerly brought extension workers into villages to assist farmers in diagnosing and remedying agronomic problems on site. Structural adjustment advisers introduced a 'demand-driven' approach as part of a wider effort to privatise government services and reduce public spending. One farmer told us, after extension officers stopped visiting farms, that "farmers were left at the mercy of the weather and God" (Interview, September 2014). In the demand-driven extension approach, farmers were required to travel to government offices and seek help when they needed it. However, we were told by participants that "farmers cannot afford to go from the village to the provincial or higher level offices due to lack of resources" (Wote FGD, 8 March 2012).

Often, when farmers did seek assistance and needed the extension officers to visit their farms, they reported having to pay for gasoline or diesel to fuel the extension officer's vehicle. Many farmers, especially women, reported feeling too intimidated to enter official government premises and request help. They were also not always aware of problems that needed attention until it was too late. Further, many farmers felt that extension workers could not protect farmers from 'middle-men,' who come to "spoil the price of our produce" and "benefit from the sweat of the farmers." One farmer noted that "the 'demand-driven' approach is problematic as most farmers do not know who to approach and the government's officers are few and not available at all times." With the obstacles they faced, most farmers were consigned to proceeding in their endeavours without relying on government assistance or advice: "We just develop ourselves," noted one (Makueni FGDs, 2011).

Another farmer in Wote described her view that a peer-to-peer agricultural extension model has the potential to fill in the gaps in the flow of agricultural information. Peer educators "will be like a bridge between the extension workers and the community. So the community will be involved, and these peer-leaders can give feedback to the extension workers. And the extension workers will be getting information from the community. So there will not be such a disconnect. There will be not such a gap. And [through the same channel] maybe this information and any change of policy will be flowing to the community" (Interview, December 2011). Some farmers also suggested that the civic education of farmers in county-level forums might enable more effective participation in policy-making processes.

These various efforts at self-development were, however, often stymied by a lack of access to resources, including finance. In most rural areas of Kenya, the very common practices of 'merry-go-rounds' and group saving societies maintain a basic level of financial security for those who participate (see Chapter 6). However, these informal arrangements and sources of finance are not always available at sufficient levels or in a timely manner. Table banking, especially among women's groups, is a 'scaled-up' form of informal financing increasingly employed by citizens, and involves similar patterns of group saving, loans and investments (Obiria, 2014). One retired agricultural officer emphasised to us the great potential of more accessible rural credit to positively impact efforts for household food security:

> *The government devolution of financial power down to the county would have in its package some money for agricultural production, and not only for agricultural extension, but also for investment in agriculture. There is in the country what we call Agricultural Finance Corporation, where people can get credit. I believe this particular organisation is going to be still available to the counties, for farmers in the counties to go and get credit to do agricultural production. Credit availability to the farm, apart from the government's investment into the sector at the county level, would be an added strength towards food production. (Interview, 2012.)*

The greatest impediment to accessing credit from banks and microfinance institutions is the requirement of collateral that in most cases is the land title deed. This situation greatly disadvantages women, as most do not own land (see Chapter 4). For those who do have title deeds, they are reluctant to use them as collateral due to fear of foreclosure in case of a default. These impediments speak to the larger equity questions that require attention, and they also provide an important indication of why so many poor Kenyans, especially women, turn to informal group savings and credit arrangements such as table banking to save money and secure small loans (Obiria, 2014; see also Chapter 6 in this collection).

Our sample of farmers also discussed the potential for establishing an 'Agriculture Stakeholders Forum' "with teeth"—or decision-making power—to get their messages across to agriculture-related policy-makers (Interview, 8 March 2012). They expressed an interest in being more involved in decision-making, and a willingness and capacity to engage directly in the exchange of and debate about relevant information through community extension, policy-making forums and other devolved county governance structures that are in the process of being established. One such space for citizen participation is detailed in the Public Finance Management Act (2012), which envisaged the establishment of County Budget and Economic Forums (CBEF) by each county government to facilitate participation of local residents in prioritising resource allocations to development, including agricultural services. These forums are not yet fully operational, with only half of the counties having constituted the forums at the time of writing. Additionally, the process of nominating non-state actor representatives remains unclear, making them susceptible to elite capture.

This is not to say that county governments have been idle since their election in 2013. Various steps have been taken across the country. For instance, each county has prioritised the top three agricultural commodities or value chains to be given preference and policy support in that county. Government revenue is up: "The taxman's collection from county authorities has gone up seven-fold since the onset of devolution thanks to growth in economic activity" (Wokabi, 2015). And in Makueni, and in many counties, talk radio shows and community meetings provide two concrete opportunities for engagement. *Barazas* (community meetings) promoting a "Grow more food" campaign in Makueni, exemplify the priority the county's agriculture department places on food security and, moreover, on widening information channels and making more room for two-way dialogue between the governed and the governors. According to Mr. Jackson Katua, Deputy Director at the Department of Agriculture, Makueni County, "The Department of Agriculture sees its central mandate as achievement of food security and efforts are being made to involve all stakeholders in achieving this" (Interview, September 2014). The shape and efficacy of these efforts await further development and study.

Farmers had other valuable ideas about how to improve information flows and two-way communication with policy-makers, including "making more farmers' groups" where "farmers can present their views and these can be forwarded to policy-makers" and greater farmer involvement in research field days and demonstrations. Other priorities included water infrastructure, credit facilities and agronomic training—including the training that could be provided by their own peers.

Discussion and conclusion

Public participation in agricultural policy-making has, to date, been less fully and evenly implemented than envisioned by Kenya's constitution (Committee for the

Implementation of the Constitution [CIC], 2014, p. 11). This is not particularly surprising, given that county officials only took office for the first time in March 2013. Key challenges include issue prioritization, inter-governmental relations, an unevenly managed transfer of devolved functions and county boundary disputes. Despite this, progress is being made. For instance, ongoing legislative reform is embracing citizens' participation and the accrual of community benefits, for instance in regard to mining ventures and revenue collection, with the nation's courts generally upholding these principles when challenged.

At a 2015 forum reflecting on two years of devolution, Professor Winnie Mitullah of University of Nairobi's Institute for Development Studies identified participation as the way ahead and "questioned how the public was being involved in forums and platforms that require their participation" (Kibet, 2015). She also pointed to the need for studies that consider why the public was not engaging in civic undertakings and called for guidelines and structures to facilitate public participation (Karanja, 11 March 2015). According to a June 2014 Transparency International survey in sixteen counties, the public was shown to lack adequate knowledge of county governance processes. When asked about citizen consultation forums, only 38% of the respondents were aware of the meetings and out of these, only 15% attended while the rest cited other commitments and a lack of interest as reasons for their non-attendance (TI, 2014). It may be pertinent to ask, following on Dr. Mitullah's suggestion, whether targeted information campaigns and other efforts to facilitate public participation can reverse the reported 'lack of interest' in civic engagement. Indeed the Committee for the Implementation of the Constitution (CIC) developed public participation guidelines (Kibet, 2015), but these are as yet only guidelines, and not policies with the full effect of the law. It is equally interesting to note that the National Parliament which County Assemblies borrow the legislative tradition from, is yet to enact public participation and access to information laws.

As food and nutrition security spurs wider human socio-economic development, similarly the progressive realisation of citizen participation contributes to the achievement of other constitutional goals. Wide and innovative public participation has the power to rein in corruption (Devas and Delay, 2006). This is reason enough to energetically facilitate public participation processes. If accomplished early, well-planned public participation can ensure that subsequent governance processes and decisions are more likely to be representative of the diverse public's views and choices, thus strengthening the foundations of the rule of law as both citizens and state share stakes in the outcomes of their devolved democratic decision-making. It would also be through the development of these democratic relationships that Kenyans could contribute to a more equal standing in negotiations with external funders and investors, such that these foreign actors cannot as readily circumvent the country's sovereign governing processes and decisions.

Kenya is clearly undergoing a complex transition including from the old to the current constitution; from the central to the national government; from local authorities to county governments; from a unicameral parliament to a bi-cameral one (Senate and National Assembly); and from a pliant judiciary to an independent one. The old relationships and attitudes that prevailed under the previous system may be more difficult to change than the structures of government.

The main findings arising from our case study that may help smooth this transition are for policy-makers concerned with devolution to prioritise mechanisms for citizen participation and then to study and plan the way forward, and fast-track civic education and information dissemination on participation in devolved governance. The media provides one efficient means of reaching people, though the greatest potential may rest in a turn towards social media, information technologies (ICTs) and radio platforms to reach and engage youth, women, men, the disabled, and populations living in more remote regions. These would help diversify citizens' access to information, echoing what one farmer stated to us in a 2012 discussion: "Information centres and communication systems [need] to be developed to upgrade farmers' information. But as of now, farmers use mobile phones to share information" (Wote, 8 March 2012). And since wider public participation builds networked, rather than top-down, processes of governance, an additional objective can be further enabled, in that citizen participation can support more robust relations of accountability.

To support this recommendation, we here turn to a consideration of how our study's findings pertain to facilitating citizen participation. For convenience, we consider these suggestions in the categories of 'who' participates, 'what' they participate in, and 'where and when' this participation takes place. The question of 'why' is answered largely by the constitution's emphasis on peoples' sovereignty. Literature on devolution around the world suggests different paths for the achievement of different forms of devolution, whether technocratic or transformational. Some further suggestions are then made as to 'how' participation might be effectively facilitated.

Who has access to devolved governance processes?

In interviews and focus groups conducted in Wote in August–September 2014, there emerged mixed sentiments as to which stakeholders should participate in agriculture-related decision-making. The majority agreed, however, on efforts to target various agricultural value chains (e.g., mango, citrus and poultry) identifying, in a participatory manner, the key stakeholders and especially encouraging the engagement of representatives of these groups.

Inclusive public participation does not only mean more participants; it also means more diversity of participants, with special attention to classes and groups historically marginalised and under-represented. The largest of these is 'the poor' in general, and women subsistence and small-scale farmers in particular. Whereas a technocratic devolution requires improved efficiency and accountability within the existing system, a transformative approach is likely to change who participates as much as to whom and to what principles decision-makers are held accountable. This deeper expression of devolution transforms accountability from a predominantly upward-facing orientation of state officials to external funders and the executive branch of government, into an orientation that gives equal weight to horizontal accountability of elected officials and downward accountability to the public. This echoes the expressed interests of those we interviewed.

What processes are citizens to join?

The constitution envisions public participation in the business of the county assemblies, such as planning and budgeting. According to Kenya's constitution, citizens have "the right to petition government, to recall elected officials, and to call a referendum (supported by a petition signed by at least 25% of the electorate)" (ICJ, 2013, p. 34). It is not only public money expenditure which is under debate in county decision-making, but also how money is sourced when it comes to county credit and debt. Devas and Delay note that "much of the original interest in decentralisation from the World Bank and regional banks stems from their desire to lend money to subnational governments. …loan finance has all too often been channelled into prestige capital projects. Debt servicing then consumes resources that could have been used for improvements to the delivery of basic services and the maintenance of existing assets" (Devas and Delay, 2006, p. 688). County debt, then, requires special attention from the public to ensure that loans, which citizens will immediately or ultimately repay, actually are spent on expenditures that in fact address citizens' needs rather than supporting prestige projects that too often become 'white elephants' (see, for example, Desai, 2015).

Where and when?

Section 113 of the County Government Act lists means of facilitating citizen participation in county governance, including social media–based platforms, town hall meetings, budget preparation and validation forums, notice boards, development project sites, and citizen fora at county levels (ICJ, 2013, p. 34–35). The guidelines formulated by the Commission on the Implementation of the Constitution build on these legal foundations to specify further where and when participation takes place, but as earlier inferred, require more solid rooting in policy and law.

Elsewhere in Africa, a variety of forums have been devised as spaces within which public participation takes place. For instance, Dauda showed how, in South Africa, the public participated in policy-making through community development forums, which included "political parties, civic organisations, business associations, churches, sporting bodies, organised labour and ratepayers or residents associations. …[with] an executive committee chosen by the participating organisations, which … became the focal point for Council policy" (Dauda, 2006, p. 298). The development of terms of reference and appointment of steering committees to help guide the participation of community groups helped improve accountability and effectiveness of these forums (Dauda, 2006, p. 299). This example might be pertinent to the development of guidelines for public participation in Kenya's counties, considering that Kenya has a rich and continuing history of self-funded community-based organising, along with widespread experience with the trust and cooperation needed to maintain these ubiquitous forms of horizontal, democratic collective action (see Chapters 1, 2 and 6 in this collection). Included in this debate will be questions about how to ensure the accountability of civil society organisations, not least to the public they serve.

How?

In their study of decentralisation of resource management policy, Rahman et al. (2012) identify *learning, information dissemination and capacity building* (equally for citizens and state officials) as central to engendering effective participation and inclusion at the community level. Devas and Delay likewise note that "local elections need to be supplemented by opportunities for more direct citizen participation in decision-making, and by greater information about the availability and use of resources. This can also help to build accountability, since those who have participated in discussions about the local issues are more likely to demand accountability" (Devas and Delay, 2006, p. 684). This underlines our argument that participation and good flows of information are critical to building accountability, especially downward accountability. "Accountability requires that both citizens and central government have accurate and accessible information about local government: about available resources, performance, service levels, budgets, accounts and other financial indicators" (Devas and Grant, 2003, p. 310–311). Section 87 of the County Government Act provides guidance for ensuring citizen participation in county governance through the provision of information. The Act's first principle is citizens' rights to "timely access to information, data, documents, and other information relevant or related to policy formulation and implementation." Other principles defined there include the "protection and promotion of the interest and rights of minorities, marginalized groups and communities and their access to relevant information" (ICJ, 2013, p. 34). Section 113 of the Act further "makes public participation in county planning processes mandatory. It also requires … the provision to the public of clear and unambiguous information on any matter under consideration in the planning process" (ICJ, 2013, p. 34). In a September 2014 interview, a Makueni County government official reported that the County Assembly had developed curriculum and engaged full time staff within the Education Department to spearhead civic education within the county. And while the current consultative process there is largely ad hoc, this is expected to change as the newly recruited civic educators put participatory structures and processes in place. Access to information and opportunities to learn are key to facilitating participation, especially among women, youth and other marginalised groups.

Next steps

Most of the early struggles with devolution experienced in Kenya are common elsewhere, and have concentrated on the process of setting up the devolved units, officials' responsibilities, functions and budgets. A proposed new agricultural policy opens the way to a different kind of test of devolution, in the form of debate over the content and direction of agricultural development, which is at the heart of development in the country and region in general. The draft policy was published in December 2014, titled "Food: Our Health, Our Security." It is the first major sector-wide agricultural policy introduced in Kenya in many years, and certainly the first since agricultural services began to be formally devolved to counties in August 2013. As such, the introduction of the "Food: Our Health, Our Security" draft policy raises for us pertinent questions about the extent to which processes of citizens' participation will impact the content and direction of debate, decisions and eventual implementation of this inaugural instance of devolved agricultural policy-making. While its vision and objectives are stated in fairly familiar terms, generally, to 'transform agriculture into a more modern and commercially viable sector' and 'mainstream food and nutrition security into the country's overall development agenda;' the means by which the policy seeks to achieve its objectives are sweeping and contentious (Wahome, 2015). The draft proposes "a carefully considered policy of reconsolidating land for agricultural production … [involving] setting apart a section or sections of a ward or location for settlement as the rest of the land is consolidated for agriculture" (GOK, 2014). In short, the policy proposes wholesale relocation of rural peoples into villages and reapportioning of their land, two processes likely to lead to intense conflict, and questionable benefits.

There promises to be vibrant debate over the provisions of the policy as it makes its way through the legislative process, especially given the deep cultural, historical and practical attachment to land that most rural Kenyans hold dear. Dissemination of information about the draft policy, along with deep, wide, diverse and persistent citizen participation in debate about it, may make the difference between a top-down policy prescription and a truly democratic devolved policy direction that offers the promise of a locally designed and publically supported agriculture policy capable of building a more food-secure future on the basis of sustainable rural livelihoods designed with the inclusion of farmers themselves.

References

Ajulu, R. (1998). Kenya's democracy experiment: The 1997 elections. *Review of African Political Economy, 25*, 275–285.

Ashby, J. A. and Sperling, L. (1995). Institutionalizing participatory, client-driven research and technology development in agriculture. *Development and Change, 26*, 753–770.

Bardhan, P. (2002). Decentralization of governance and development. *Journal of Economic Perspectives, 16*, 185–205.

Brownhill, L. (2009). *Land, Food, Freedom: Struggles for the Gendered Commons in Kenya, 1870–2007*. Trenton, NJ: Africa World Press.

Brownhill, L. and Hickey, G. M. (2012). Using interview triads to understand the barriers to effective food security policy in Kenya: A case study application. *Food Security, 4(3)*, 369–380. doi:10.1007/s12571-012-0183-2

Chambers, R., and Jiggins, J. (1987). Agricultural research for resource-poor farmers Part I: Transfer-of-technology and farming systems research. *Agricultural Administration and Extension, 27*, 35–52.

Commission for the Implementation of the Constitution (CIC). (2014). *Assessment of the Implementation of the System of Devolved Government: From Steps to Strides* (June Report). Nairobi: CIC.

Cornwall, A. (2008). Unpacking 'participation': Models, meanings and practices. *Community Development Journal, 43*, 269–283.

Crawford, G. (2009). Making democracy a reality? The politics of decentralisation and the limits to local democracy in Ghana. *Journal of Contemporary African Studies, 27*(1), 57–83.

Dauda, C. L. (2006). Democracy and decentralisation: Local politics, marginalisation and political accountability in Uganda and South Africa. *Public Administration and Development, 26*, 291–302.

Desai, A. (2015). Of Faustian pacts and mega-projects: The politics and economics of the port expansion in the south basin of Durban, South Africa. *Capitalism Nature Socialism, 26*(1), 18–32.

Devas, N. and Delay, S. (2006). Local democracy and the challenges of decentralising the state: An international perspective. *Local Government Studies, 32*, 677–695. doi:10.1080/03003930600896293

Devas, N. and Grant, U. (2003). Local government decision-making—Citizen participation and local accountability: Some evidence from Kenya and Uganda. *Public Administration and Development, 23*, 307–316.

Eidt, C. M., Hickey, G. M., and Curtis, M. A. (2012). Knowledge integration and the adoption of new agricultural technologies: Kenyan perspectives. *Food Security, 4*(3), 355–367. doi:10.1007/s12571-012-0175-2.

Government of Kenya. (2011). *Final Report of the Taskforce on Devolved Government, Vvolume I: A Report on the Implementation of Devolved Government in Kenya*. Nairobi: Office of the Deputy Prime Minister and Ministry of Local Government.

Government of Kenya. (2014). *Food: Our Health, Our Security*. (Draft Agricultural Policy). Nairobi: Ministry of Agriculture, Livestock and Fisheries.

Hall, A. J. and Yoganand, B. (2004). New institutional arrangements in agricultural research and development in Africa: Concepts and case studies. In A. J. Hall, B. Yoganand, R.V. Sulaiman, R. S. Rajeswari, P. C. Shambu, G. C. Naik, and N. G. Clark. (Eds.), *Innovations in Innovation: Reflections on Partnership, Institutions and Learning* (Chapter 4). Patancheru 502 324, Andhra Pradesh, India, and Library Avenue, Pusa, New Delhi 110 012, India: Crop Post-Harvest Programme (CPHP), South Asia, International Crops Research Institute for the Semi-Arid Tropics (ICRISAT), and National Centre for Agricultural Economics and Policy Research (NCAP).

Hickey S. and Mohan, G. (2005). Relocating participation within a radical politics of development. *Development and Change, 36*, 237–262.

Kantai, P. (2013). Kenya's devolution revolution. *Africa Report*, July edition. Retrieved from http://www.theafricareport.com/East-Horn-Africa/kenyas-devolution-revolution.html

Karanja, S. (2015, March 11). Majority 'back referendum on county cash.' *Daily Nation* (Nairobi). Retrieved from http:// http://www.nation.co.ke/news/Referendum-County-Funds-Revenue-Devolution-Afrobarometer/-/1056/2650220/-/4tcjbl/-/index.html

Kerrow, B. (2015, March 1). Deal with teething problems as we mark two years of devolution. *The Standard* (Nairobi). Retrieved from http://www.standardmedia.co.ke/article/2000153260/deal-with-teething-problems-as-we-mark-two-years-of-devolution

Kibet, L. (2015, March 12). 80 pc of Kenyans don't participate in county issues. *The Standard* (Nairobi). Retrieved from http:// http://www.standardmedia.co.ke/article/2000154485/80pc-of-kenyans-don-t-participate-in-county-issues.

Larson, A. M. and Soto, F. (2008). Decentralization of natural resource governance regimes. *Annual Review Environmental Resources, 33*, 213–239.

Mutunga, W. (1999). *Constitution-making from the middle: Civil society and transition politics in Kenya, 1992–1997*. Nairobi: SAREAT.

Neef, A. and Neubert, D. (2011). Stakeholder participation in agricultural research projects: A conceptual framework for reflection and decision making. *Agriculture and Human Values, 28*, 179–194.

Nyong'o, A. (2015, March 1). Develop legal mechanism for sending CDF billions to serve the counties. *The Standard* (Nairobi). Retrieved from http://www .standardmedia.co.ke/article/2000153262/develop-legal-mechanism-for-sending-cdf-billions-to-serve-the-counties

Obiria, M. (2014, November 14). Table banking lends Kenya's women the means to beat the poverty trap. *Guardian*. Retrieved from http://www.theguardian.com/ global-development/2014/nov/14/table-banking-kenya-women-poverty

Pretty, J. (1995). Participatory learning for sustainable agriculture. *World Development, 23,* 1247–1263.

Rahman, H. M. Tuihedur, Hickey, G. M., and Sarker, S. K. (2012). A framework for evaluating collective action and informal institutional dynamics under a resource management policy of decentralization. *Ecological Economics, 83,* 32–41.

Ribot, J. (2002). African decentralization: Local actors, powers and accountability. (United Nations Research Institute for Social Development (UNRISD) Programme on Democracy, Governance and Human Rights, Paper No. 8). Geneva: UNRISD. Retrieved from http://www.unrisd.org/unrisd/website/document.nsf/0/3345ac67e6 875754c1256d12003e6c95/$FILE/ribot.pdf

Rondinelli, D. A. (1980). Government decentralization in comparative perspective: Theory and practice in developing countries. *International Review of Administrative Sciences, 47,* 133–145.

Sanginga, P. C., Chitsike, C. A., Njuki, J., Kaaria, S., and Kanzikwera, R. (2007). Enhanced learning from multi-stakeholder partnerships: Lessons from the Enabling Rural Innovation in Africa programme. *Natural Resources Forum, 31,* 273–285.

Shayo, E. H., Mboera, L., and Blystad, A. (2013). Stakeholders' participation in planning and priority setting in the context of a decentralised health care system: The case of prevention of mother to child transmission of HIV programme in Tanzania. *BMC Health Services Research, 13.* 273. doi:10.1186/1472-6963-13-273, pp. 1–12. Retrieved from http://www.biomedcentral.com/1472-6963/13/273

Stein, H. and Nissanke, M. (1999). Structural adjustment and the African crisis: A theoretical appraisal. *Eastern Economic Journal, 25,* 399–420.

Sumner, A., Acosta, M. A., Kapur, R., Bahadur, A., and Bobde, S. (2008). *Access to Governance and Policy Processes: What Enables the Participation of the Rural Poor?* Rome: IFAD, Rural Poverty Report.

Transparency International (TI). (2014). *Is it My Business? A National Opinion Poll on Devolution and Governance in Kenya.* Nairobi: Transparency International, Kenya.

Wachira, M. (2013, September 28). Shallow understanding of problems devolution was to solve could lead to its failure. *East African.* Retrieved from http://www.theeastafrican. co.ke/news/Why-devolution-will-not-end-inequality-in-Kenya/-/2558/2010884/-/ upb27xz/-/index.html

Wahome, M. (2015, March 3). Why you could soon leave your land for town estates. *Daily Nation.* Retrieved from http://www.nation.co.ke/lifestyle/smartcompany/Why-you-could-soon-leave-your-farm/-/1226/2640864/-/dmw225/-/index.html

White, S. C. (1996). Depoliticising development: The uses and abuses of participation. *Development in Practice. 6*(1), 6–15.

Wokabi, C. (2015, February 28). Counties remit Sh1.4bn monthly, earning rare praise from taxman. *Daily Nation* (Nairobi). Retrieved http://www.nation.co.ke/business/Counties-remit-Sh14bn-monthly-earning-rare-praise-from-taxman/-/996/2638652/-/ if5830z/-/index.html

10 The resilience umbrella

A conceptual tool for building gendered resilience in agricultural research, practice and policy

Leigh Brownhill and Esther M. Njuguna

Earlier in this collection, we identified four constituent elements of semi-arid farming systems in Eastern Kenya that signified, for us, key sites of gendered and generational negotiation and change, in *access, resources, livelihood, and policy and institutional relations*. These instantiate the array of economic, social, ecological and political relationships that shape the *social ecologies* of local food and agriculture systems. In the counties of Machakos, Makueni and Tharaka-Nithi, majority of farmers are small-scale and subsistence-oriented. Some of the world's most vulnerable populations reside in Eastern Kenya. The intractability of the hunger crisis there appears despite the fact that for good reason (historical settlement patterns, ecological fragility, cultural cosmopolitanism and proximity to Nairobi), the region has brought generations of agricultural researchers to grapple with the region's chronic food and nutrition insecurity. As scientists seeking resolution to the hunger problem in eastern Kenya, we had to try to account for this paradox: so much research, yet still so much hunger. So we included questions of the effectiveness of the institutional relations which should link farmers, scientists and policy-makers (and markets and other institutional actors) and bring local research results to bear more effectively on high-level policy and on-the-ground practice. These policy and institutional relations, though often physically removed from the villages of eastern Kenya, are taken here as especially important forces within the local farming system, insofar as they shape the larger political economy context within which all other elements of the system operate.

To more holistically measure the complex resilience capacities and challenges within this system, we needed an conceptual-methodological tool which could integrate analyses of diverse agronomic, ecological, sociological, institutional and other practices, processes and relationships, while also being clear and 'user-friendly' enough to help us clarify, order and answer, rather than obfuscate, the complex question of how to end household food and nutrition insecurity. Finding no tool that suited the precise needs of our highly interdisciplinary project, we designed our own, using a metaphorical device—the image of an umbrella—to better capture the workings of a complex gendered farming system. We turn, below, to a description of the *Resilience Umbrella*, before employing the tool to analyse the results, impacts and implications of the KALRO-McGill project for participating farmers groups and households. The chapter then concludes with an assessment of the tool and some of the questions the research raised for further research.

The 'Resilience Umbrella'

We adopted the umbrella metaphor for gendered food security resilience because umbrellas shelter people from the weather. Umbrellas provide users with a mobile *canopy*. The canopy is given shape by interlocked mechanisms which together make the umbrella function. Our conceptualization of the umbrella in analytical terms posits the canopy as representing *food security resilience*, and the umbrella's constituent mechanisms, the four sites of gendered power, negotiation and chance in relations of access, resources, livelihoods and policies (see Figure 10.1).

Users hold the *handle* to employ the protective services of the umbrella. The *handle* represents access and entitlements, or the means through which household members secure and maintain rights to resources. The handle is connected to the umbrella's *shaft*. The *shaft* holds everything together, and so for us, stood for the diverse natural, social, financial and other resources that constitute the productive assets of the farming system. The *ribs* hold up the canopy. The *ribs* are the livelihood strategies and activities that are undertaken by farmers as individuals and households, in groups, networks and communities. These activities give specific shape and strength to the farming system. The *stretchers* are connected to the shaft, and are the key mechanism that activates the ribs, and therein, opens the canopy itself. We see the *stretchers* as the policies and institutions (including state, non-state and market) that on the one hand, hold on to, or have significant power over, many of

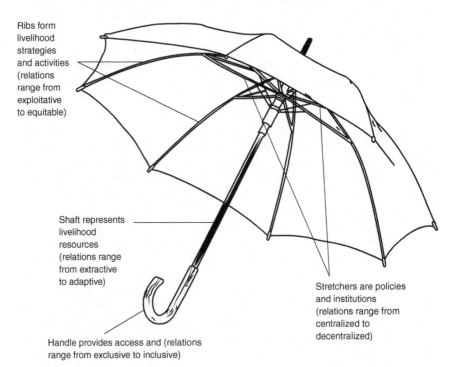

Ribs form livelihood strategies and activities (relations range from exploitative to equitable)

Shaft represents livelihood resources (relations range from extractive to adaptive)

Stretchers are policies and institutions (relations range from centralized to decentralized)

Handle provides access and (relations range from exclusive to inclusive)

Figure 10.1 Resilience umbrella.

the farming system's resources and on the other hand, can shape farmers' access to those resources to affirm and support farmers' livelihood strategies and options.

In this way, the different mechanisms make up the whole system, or umbrella. But what is an umbrella actually? The canopy is a flimsy piece of cloth that can do little to protect its user, unless its simple but critical ribs, stretchers, shaft and handle are present and operational. Likewise, we propose that food security resilience is constituted by the coordinated operations of socially and ecologically resilient livelihood activities, policies, natural and social resources and mechanisms for equitable access. Like an umbrella, food security resilience is not an abstract idea; it is something made by people and held (or not) by people. Unlike umbrellas, resilience cannot be mass produced in factories and purchased in shops; resilience has to be created by those who use it. The changes wrought by building socio-ecological resilience in the farming system are here understood to be mainly relational—more equitable inclusion, diversity, participation, learning, multi-level development, and enhancement of local knowledge.

The Resilience Umbrella is not used here in the same way that it is commonly used in the insurance industry to symbolized assurance against risk and loss. If we borrow the idea of the three *nested economies* discussed in Chapter 6, an insurance umbrella is intended to cover the market economy, while our resilience umbrella, at least to begin with, covers the subsistence economy. In fact, though, farming system resilience is here taken to be adducible at multiple levels, from individual, household, farming-system and county, to national, regional, continental and global scales. The Resilience Umbrella as an analytical lens is adjustable, to take in one or more elements, at wider or more-focused scales. Finally, the umbrella tool not only identifies the parts of the system, but considers the relations and practices within each part on a comparative scale, as follows:

1   Access and entitlements, relations range from exclusive to inclusive
2   Livelihood resources, relations range from extractive to adaptive
3   Livelihood activities and strategies, relations range from exploitative and equitable
4   Policies and institutions, relations range from centralized to decentralized

The gender transformative scale, introduced in Chapter 1, offers a means by which to use these *ranges of relationships* to help us gauge the direction that a project, or a farming system, is moving, or where it proposes to go and where it reaches, as assessed at a given time. Bahadur, Ibrahim and Tanner's (2013) ten characteristics of resilience provide entry points for assessing those relationships, and thus determining direction of change.

Our Resilience Umbrella, when the social relations (in resilience terms) of all of its elements (access, resources, livelihoods and policies) are considered, can show how the farming system is tilted towards exploitative relations, which leave women out in the rain, so to speak, or towards transformative relations, which shelter men, women and children from hunger (see Figure 10.2). Transformative relations are measured in the Resilience Umbrella frame not only by degree of equitable access

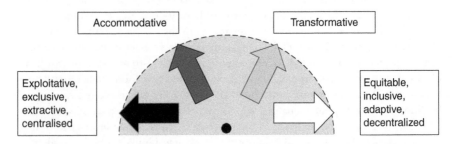

Figure 10.2 Gender-transformative scale.

to resources and livelihoods, but more importantly, the degree to which this gender equity strengthens household food and nutrition, especially for children (Sanginga, Waters-Bayer, Kaaria, Njuki, and Wettasinha, 2009) as well as ecological well-being. It is this last critical point—the contribution of gender equity to food security and farm system resilience—that for us justifies, and even demands, wider policies and institutional support for gender equity in law and in practice across sectors and governance scales. We move to that analysis below.

The Innovating for Resilient Farming Systems (INREF) project's practical and action-oriented development research resulted in significant gains in gender equity and strides towards food security resilience by farmers participating in the project. In the next section, some of these results are considered, in particular with reference to the character of the social relations of access, resource use, livelihoods, and policies.

Resilience results of the project

From its inception, the INREF project identified with the view that the empowerment of women is a prerequisite to the attainment of enduring food security. With this commitment, and in recognition of the central role they play in semi-arid farming systems, we designated women's and mixed-gender farmer groups as the primary participants in the project. All of the project's multi-disciplinary research and field activities were concentrated around these groups' farms and their shared nutritional, economic, ecological and agronomic activities and concerns (see Chapter 2). In addition, many researchers in the project team included 'non-participating' farmers, farmer groups and communities, whether in comparative analyses or in order to consider wider socio-cultural context and political economy relations.

In the 54 farmer groups that ended up formally participating in the project over its 3.5-year tenure, women's total membership stood at 66%. This starting point facilitated inclusion of women's often-overlooked indigenous knowledge and a recognition of the sometimes-hidden priorities that women farmers maintain when making farm-level decisions, including what have been called elsewhere in this collection, *non-priced values* (see Chapter 7). In the design of the project's crop evaluations and other activities, group members were positioned not as research subjects but more as students, field evaluators, peer teachers, local experts, and indigenous innovators. We now turn to a consideration of results from these activities.

Access and entitlements

Multiple information-networking models (e.g., for peer-to-peer education) were employed to improve participants' dietary diversity, livestock management, and post-harvest crop handling, among other capacities. Each was linked to the crops being evaluated in the field trials. Representatives from the farmers' groups were trained as peer educators and service providers, and in turn reached out to their groups and larger networks with peer education and service provision. The research objectives of these educational activities included investigating alternative agricultural extension methods; examining the extent to which researchers' exogenous knowledge, technologies and practices were adopted and adapted by farmers; and the degree to which the resulting syncretism of knowledge helped the farming system to be more resilient, or better adapted to changing environmental and social conditions (see Chapter 2).

An innovative information pathway followed on from farmers obtaining rainfall data across the research sites. While the government's meteorology department maintains rain gauges in the counties, these were situated very far from the farmers' villages. Even if they could have accessed the rainfall data from the internet or radio, results were not as applicable as if they had been obtained from the farmers' own agro-ecological niches. So, the project's soil team provided 18 rain gauges (one to each set of three farmer groups). Groups identified one farmer to take the daily readings. Most of those who volunteered to take care of the rain gauges were women. They were keen to use the new tools; and they were always around the homestead where the gauges were installed. They were trained to measure and record the rainfall data at the same time every day, to empty the gauges and reset them for use.

One research team in Machakos County came up with an innovative way of interpreting the rainfall data. Based on farmers' recall and estimates of previous seasons, the farmers and researchers together determined that the total average amount of rainfall received in the area over the years was approximately 300mm per season. They recognized that not all crops would grow well with that little rain. So if the farmers were to make the best out of the rain that they were likely to receive, then dry planting before the rains started would be best, especially for maize and beans. As the days went by and rainfall data were tracked, if the amount received was less than expected for that time, then farmers could opt to plant sorghum, green grams and cowpeas, which are able to grow with less water.

With this in mind, as the planting season approached, the women farmers in one village started passing through the home of the farmer with the rain gauge, where they would ask: 'How much has it rained so far? So how much rain is remaining for this season? That means I am late to plant maize now? Is there still time to plant cowpeas?' Women engaged in this innovative learning and networking activity increased their resilience by diversifying their access to information, and using local community knowledge to build preparedness in the face of uncertain seasonal rains.

Based on their analysis of the rainfall data, farmers were empowered to make well-informed choices as to what to plant and when to invest in inputs for their crops.

Women small scale and subsistence farmers, and many men as well, are excluded from access to many resources, including education, as seen in the generally low literacy rates, especially among women. The example of the simple, farm-based technology of rain gauges helps us show that more inclusive knowledge-sharing can spark innovative action to improve farming and food security outcomes, despite the presence or continuation of other forms of exclusion.

Studies in this collection show that *shared values and social networks* matter in building farm system resilience, importantly in *diversifying* access to information. Networks and the density of their information pathways help farmers also to *accept and expect change and uncertainty* and therefore to *plan and prepare* in advance. Chapter 3 underlined the significance of trusted and *effective leadership* as a positive force in rural communities, where their good examples of service to the community make farmers more willing to *learn, participate* and adapt.

Livelihood resources

Women farmers' collective action is historically rooted and remains richly textured in the 21st century in Kenya. If their action in groups and networks give subsistence and small-scale farmers access to more information, it equally provides access to other resources, including labour, land and small-business investments (See Chapters 3 and 6 in this collection). Kenyans generally congregate in high numbers in social groupings of all kinds from registered to informal organizations, networks, alliances, movements, commissions, trusts, jointly owned businesses, and more. The variety of areas of life that Kenyans attend to with some form of collective social organization is, to outsiders, often astounding.

Collective action by farmers is therefore an important form of social capital in Kenya's rural communities in the semi-arid counties. However, not all collective action is created equal: A farmer group can be generated by farmers themselves or established by researchers or private interests for the purpose of experimentation or labour and resource extraction. Collective labour can be organized from above, and tied in to *extractive value chains* (focused on channeling resources and profits out of farming communities). On the other hand, autonomous collective action, organized from below by diverse farmer groups, can be mobilized for purposes set by those groups' members, including the shaping of *adaptive value chains* (focused on use of resources for socio-ecological resilience and food security within farming communities). We worked mainly with autonomously organized farmers groups in research aimed at building adaptive capacities and building resilience in the farming system. Some of the results of our efforts are reviewed here.

Groups' collective action

In many of the participating groups, women took responsibility for leading tours of the learning farms during end-of-season field days. They explained to visitors the attributes of the crop varieties, the natural resource management options they followed, challenges they encountered, how they solved them and the results they obtained in their overall efforts. Farmers saw results when other group members and neighbours started introducing the same innovations in their own farms. Some groups collectively raised funds to purchase seeds and fertilizers. They were sure of the fertilizer type to use, which was informed by soil analysis results from their learning farms, and when to apply it, before the start of the rains.

Those who took up peer-education roles generally gained confidence in their abilities and in the results they had obtained. They were often especially proud of their farms' high yields, and were able to compare their success with neighbours who had not adopted similar practices, and whose crops had sometimes failed. In these situations, clearly the weather was the same for both farmers; and if they had not adopted the improved practices, their crops too would have failed. The changes that led farmers to achieve these yields included the soil water harvesting techniques used, the seed varieties they selected, the date of planting, and the addition of soil amendments (manure and/or fertilizer). Their results, in turn, generated substantial interest from the community, local radio programmes and the local administrative and political leadership, which further encouraged farmers to continue their efforts in the next season.

Farmers' seed varieties

Women are the main custodians of dryland cereals and grain legumes seeds in Kenya, especially in the very remote and rural locations of Tharaka-Nithi and Makueni. Private seed companies hardly extend to such remote areas because of high transport costs, generally low consumer purchasing power and low returns on investment. Women in the semi-arid counties, like elsewhere, are also the main decision-makers when it comes to household diets. But while they are most responsible for the family's food provisioning (production or purchase, processing, preparation and presentation), they are not always in charge of all the resources needed to best fulfill this role, including land and other resources. Seed, especially food crops of local indigenous varieties, is one resource that is very often held under women's control (see Chapter 5). Household gender inequalities reduce women's capacity to access resources like land, but also reduce their use of the resources they do have in their control, such as local seed varieties. For seed must be planted in the land; and most land is owned and controlled by men (see Chapter 4), who may or may not allocate sufficient land for women's subsistence crop preferences. Biases in science and policy have reinforced these inequalities, for instance by too often sidelining agricultural research and service provision related to the indigenous seed varieties that women prioritize and prefer, and instead concentrating on export crops for global markets (Elson, 1991; Meinzen-Dick, Quisumbing, and Behrman, 2014; Staudt, 1978, is still relevant).

With this in mind, it is pertinent to consider project findings with regard to gender differences in farmers' priority crop enterprises. The women farmers who participated in the INREF project strengthened their household food sufficiency and the resilience of the larger farming system, in part by their ranking of priority technologies for inclusion in the research and on-farm evaluations. Their indigenous knowledge was respected, as their seed varieties were compared alongside the new varieties produced by the KALRO research stations. Where farmers' seeds outperformed the improved varieties, this was acknowledged, and these local varieties became part of the researchers' recommendations to other farmers. With time and resources, more could have been done, for instance in marketing these women-preferred varieties, as we discuss in the section below on livelihoods.

Before turning to that, consider this interesting process that revealed itself during the course of the initial stages of the research in 2011 and 2012. Researchers and farmers first sat together to discuss the types of crops to be included in the research. Because local farmer seed varieties were being comparatively evaluated against KALRO varieties, the histories of those local varieties were sometimes discussed. This is how the researchers found out that the seed varieties that farmers called 'local' included seeds that were, upon laboratory analysis, shown to be descendants of seed varieties produced by KALRO and disseminated in the 1970s and 1980s. These were seed varieties that had survived in this particular area for almost 40 years because they were able to grow within the local ecological conditions and farm input constraints. Those older improved varieties were adopted and deeply integrated into farmers' low-input farming system. This example showed us that the success of farmers' adoption of new seed varieties or new farming technologies and practices depends as much on how well farmers can adjust their methods to accommodate the change, as it does on how well the new varieties can adapt to the ecological and social settings in which they are being accepted.

Seed is of major concern to any effort towards the resilience of agriculture and adaptation to economic, environmental and social change. Eastern Kenyan women's indigenous seed knowledge, germplasm, and food crop priorities provide an invaluable source of agricultural adaptation potential. A note of caution in this observation should be sounded: adaptation can readily be undermined if indigenous seeds are, as they have been in many places in the world, extracted from local communities to be patented by seed conglomerates. Women farmers have not protected, saved and innovated on their indigenous seed knowledge for countless generations, only for seed companies to 'discover' and privatize these varieties and reap the profits for foreign shareholders. Farmers have saved seed in order that the seeds' nutritional, ecological and very local market values can benefit the communities that generated this knowledge, and where these values are needed most.

In this collection's pages, contributors discuss the degree to which *less-than-effective governance* of land tenure, financial resources and seed genetic material leaves many women small-scale and subsistence farmers vulnerable to impoverishment and possible landlessness, lack of access to capital and sidelining or enclosure of indigenous technical knowledge and seed varieties. These vulnerabilities at the household level are reflected in inadequacies of resilience at farm-system level, which has wider reverberations for food security. In contrast, local strategies for *equitable* access to livelihood resources like land, finance and seeds, valorize and preserve women's *local historical, cultural and knowledge-based practices* and social relations, including female husbands (and other tenure innovations), merry-go-rounds and matrilineal seed systems. Policies and institutions concerned with use of resources for farming might be made more effective, if they gave *more equitable consideration and support* to farmers' *local knowledge and practices* which have for so long served them well in their efforts towards food security resilience.

Livelihood strategies

By the end of the second season of working with farmers, many men and women farmers started seeing yield and nutritional improvements and asking an important question about next steps: '*We have been learning how to manage our crops; we understand that we can improve our yields by applying manure or fertilizer and harvesting rainwater; what, then, shall we do with our surplus production?* This question preoccupied the research team because we had witnessed many other projects get to the point of improved yields and crop performance, only to find that farmers could not access markets. This has been a long-standing disincentive to farmers' investment in improved crop production, especially given the cost of these improvements. This was one of the constraints we aimed to overcome when designing the agro-enterprise development activities of the project.

To address the farmers' question, and the project's objective on linking farmers to local input and output markets, we worked with farmers to develop participatory markets in locations as close to farmers as possible. To briefly recap what Chapter 2 covers in detail: The project's agro-enterprise development team partnered with an NGO called Cascade International to offer farmers training through an approach called 'Participatory Market Systems Development.' The training entailed clusters of three farmer groups working together to constitute a committee to spearhead a group-marketing venture. These committees were called *marketing opportunity groups* (MOGs). We adopted the progressive 'two-thirds gender rule' of the Kenyan constitution to ensure that women comprised at least one-third of the membership of these committees. In the event, the 18 MOGs that were formed achieved a women-to-men ratio of 2:1.

The committees' first responsibility was to assess local trading centres where the community usually went to buy farm inputs and sell their produce. There, they mapped the market's various value chains, and ranked the product that most farmers in the group could and would want to produce. For most of the MOGs in the three counties, green gram was identified as the crop of choice. The MOGs then identified retailers, aggregators or transporters that could buy their crops. They researched to identify the standards expected by the buyer, what quantities would be needed, what were the terms of sale and how much they could negotiate the price.

In most of Eastern Kenya's trading centres, there is usually a 'cereal stockist' shop. Most of those shops buy produce from farmers in small quantities. They normally offer very low prices for a kilogram of produce because it takes a long time for them to accumulate the large quantities they need. They generally invest in grain storage protection measures to keep the produce for a number of months, before selling it back to the community at a handsome profit, usually after a long dry spell. Some of them sell on the grain to bigger markets or even for export. Under these conditions, the stockists were found to be willing to engage with the farmer groups who promised to deliver grain of high quality in large quantities. Farmers were therefore able to offload their surplus grain as soon as they harvested.

Farmers were encouraged to assess their closest trading centre because this both strengthened local value chains and at the same time increased women farmers' participation in marketing activities. As the groups learned about engaging with the markets, they began to make group decisions and commitments concerning what to grow, how much to grow, and to whom to sell under what terms. While the project provided training, it was not involved in any kind of contractual obligations or market transactions on farmers' behalf. They were free to decide if they wanted to engage with the buyers or not. All of these marketing activities were implemented during the dry season to ensure that the farmers had time to plan and plant for the market.

The MOG in Yatta, Machakos County, set a target amount of green grams that they wanted to produce, then assessed each member's land and determined how much each was able to produce. They pooled money to purchase green gram seed and fertilizers, and went through a season of planting, weeding and harvesting their crop. They made appointments with the buyer, who sent a truck to pick up the group's produce from one of the farms. After they were paid a premium for their wholesale produce, they started receiving calls from other traders asking if they had more green grams to sell. One of the farmers expressed his surprise at the change in relations: "In my life, this is the first time we have been able to negotiate with a buyer from a point of power; usually they give us the prices they want, now we can negotiate for the price we want." Another remarked, "It is incredible, we have always complained we don't have markets for our produce, yet the buyers are looking for big quantities."

Using the participatory market development approach, farmers in the Yatta MOG earned almost Ksh 1m (approx 12,000 USD) from the sale of green grams. Women in the group reported pooling some of their share of their earnings in their women's groups. Some groups focused on purchasing household utensils for their members; some built two-bedroom earth houses; others invested in school fees for both children and grandchildren. For many of the women, this was the first time they had earned enough money from their farms to readily afford such expenses.

It has been widely documented that when agricultural markets improve, women get edged out of gainful nodes of the value chain (Gurung, Ssendiwala, and Waithaka, 2011). This happens for instance if the marketing process requires women to travel a long distance to markets to deliver their produce. When market outlets are far, women start negotiating for others to deliver the produce for them. Usually such decisions relate to cost, cultural and time constraints that most women face (Ragasa, Sengupta, Osorio, Haddad, and Mathieson, 2014, p. 11). If markets are too far, and if delivery arrangements cannot be satisfactorily made, then women often decide not to engage with market value chains at all and make a conscious decision to produce only for household subsistence purposes. And while this may serve immediate household food needs, it may also dissuade male farmers from allocating land sufficient for women's surplus production of those crops, which, because they are well adapted ecologically, could otherwise serve as an alternative in case of failure of other crops.

Market inaccessibility and exploitation by some brokers and middlemen also negatively affect men. As one farmer from Machakos told us, "We have big farms that we don't cultivate for fear of surplus produce." And though the MOGs in the project chose to focus on a crop that is generally more often prioritized by men than women for its cash-generating qualities, the participatory marketing model could be adjusted to focus on women-preferred crops too. The same model could be used to bring into focus the very local markets that women can more readily access (at farmgate, at farmer-organized auctions, at weekly village markets) and subsistence value chains (such as in high value traditional crops[1]) which could undergird support for women farmers' production of surpluses of crops that equally serve household food sufficiency.

One Yatta women's group decided to try the participatory marketing model with their own preferred commodity: indigenous chicken. Indigenous chickens are widely a considered a women's enterprise (see Chapter 7). A woman can raise and sell chicken in many cultures in Kenya, without having to account to their menfolk for the money they earn. In the Yatta case, instead of each woman selling one or two birds whenever they had a need, they decided to synchronise their sales as a group. On a day of their choosing, they brought their birds to a central location, usually the home of the group's chairlady, and invited buyers to attend the sale and bid for the birds. With this method, they were able to raise the average selling price of a cock from 250 to 400sh.

The *learning* that took place in the market development activities was not confined to farmers: researchers learned from the farmers about women's subsistence priorities, preferences and expertise, and also about the challenges of more widely developing market participation for women who might wish to trade in their preferred high-value traditional crops and indigenous livestock. Our vision of more *equitable access* to livelihood strategies involving markets suggests not only bringing women to the value chains, but bringing the value chains to women, as the Yatta MOG did when they bought buyers to their farms. This could be further realized by conceptually recognizing the non-priced (nutritional, ecological, subsistence-oriented) values associated with women farmers' prioritized food crops; and by practically exploring *multi-level perspectives* on markets, including the very local value chains that can overcome women's market constraints and strengthen their capacities to satisfy nutritional and ecological needs.

Through these project results, many participants began to make markets work for small-scale farmers by using market opportunity groups to enter their local market systems. Having done so, the question remains as to whether past patterns of exclusion and exploitation will be maintained, and undermine women's household subsistence efforts, or be transformed by the men and women of these localities, so that markets are more equitably shared. Women and men need not be in competition for limited market space and market niches—we have elsewhere discussed the debate about men's crops and women's crops, and there concluded that few clear-cut lines separate the two (Njuguna, Brownhill, Kihoro, Muhammad, and Hickey, in press).

For subsistence producers, traditional and local food crops are often not the most commercially attractive varieties because they are demanded only by customers within very local market niches, or only by the poor. But these foods are often preferred by women farmers because they are delicious, nutritious, easy to cook and locally adapted to grow in their agro-ecological zone. Making markets more accessible for women, then, could include linking these short value chains to connect women farmers of traditional crops with poor mothers who need low-priced nutritious foods for their families available in local markets.

Policies and institutions

Models of farmer-led innovation and peer-to-peer extension (in nutrition, animal health and crop health) emerged as positive and practical outcomes of the project and highlight how the INREF approach can align with county-level agricultural extension service needs. The farmer-led approach gained the attention of Ministry of Agriculture officials in the three counties, where they proposed it be used to supplement their county extension activities and initiatives.

After the implementation of structural adjustment programmes (SAPs) in the 1980s, the Kenyan government came under considerable pressure to scale down its dominant role in the national economy. Kenya's agricultural extension budget together with extension staff numbers has plummeted significantly since the structural adjustment programmes, resulting in very high ratios of farmers to agricultural extension officers (FAO, 2013). While the FAO has recommended a ratio of one extension officer to every 400 farmers, Kenya's ratio is more like 1:1500 (Akuku, Makini, Wasilwa, Makelo, and Kamau, 2014). To cope with this challenge, in the early 2000s, agricultural extension officers took up the practice of working with farmer groups rather than training and visiting individual farmers (Government of Kenya, 2011). This has led to an interesting and perhaps unexpected development within the past decade. More and more men and youth in rural farming communities are forming and joining groups for diverse activities, from farming to business and social pursuits. Men and youth have even joined women's groups. These groups provide farmers and other stakeholders with bases for diverse learning opportunities. As noted earlier, groups are also sites of shared labour and other resources.

During farmer field days in the region, farmer groups shared not only information about crop production and agronomic issues. Women and men farmers also set up nutrition tents, where they shared information on nutrition based on their own food types and typical local dietary composition. If more agricultural development and research projects included a nutrition component, farmers would be enabled to consider nutrition security for the household while making production and value chain engagement decisions. The approach of training peer educators and service providers from among the farmers is very practical in that it ensures that knowledge and innovations remain resident and circulating within the community long after time-bound research projects are over.

Beyond agricultural extension and service provision, the research findings also shed light on wider processes of setting of standards and norms for agricultural policy-making. Here, the farming-system resilience framing offers guidance in the form of principles and practices to bolster the social and ecological bases of the agricultural sector as a whole. These include the promotion of greater diversity of citizen *participation* in policy-making (see Chapter 9); advancement of *equitable* land tenure legislation and rights to forestry products (see Chapter 4 and 8); support for a *diversity* of formal and informal financial tools (see Chapter 6), including lowering barriers to women and youth's access to the interest-free *Uwezo* loan programmes; recognition and valorization of the vital sovereign sources of existing *indigenous technical knowledge*, seed varieties and matrilineal seed systems (see Chapter 5); and promotion of opportunities for farmer *networking, learning* and information-sharing (see Chapter 3).

In sum, our findings point to the need for an overarching paradigm shift away from the view that subsistence should be destroyed (see Seavoy, 2000), and towards the recognition of the market's interdependence on nature and on household subsistence political economies. For us, further research on the history and development of very local markets; the diversely priced and non-priced values of *high value traditional crops* and the dynamics of producer-consumer relations in short subsistence value chains hold tremendous promise for innovative inquiry and for informing resilience-focused policy and practical application in the future.

Ending hunger is a question of power: power to decide, to access and allocate resources, to sustain livelihoods and to find support for one's priorities. This power is held in many hands, and is contested in many domains at every level, from household to county to nation, region and world. Political leaders in East Africa, however, have a key role to play in ending hunger and malnutrition through policy formulation and service provision (Kenyatta, 2015). This goal, we have tried to show, could be advanced through the setting of standards and norms to guide policies and programmes that prioritize household food security, at the heart of which are women's and children's nutrition goals (Bukania et al., 2014; FAO, 2011). The current implementation of devolved governance is mobilizing a process of decentralization of decision-making, and holds great potential for the transformation of policy frameworks and designs to reflect the food and nutrition security goals and socio-ecological resilience of small-scale and subsistence farmers.

Eastern Kenya provides an important window onto the strengths of farmers' local subsistence knowledge, capacities and practices (e.g., social networking, land tenure, finance, seeds, livelihoods, animal breeds, forestry practices), as well as the challenges which hamper efforts to employ these capacities for food-security purposes. Perhaps the most important lesson for policy from our analysis is the need for recognition of the *non-equilibrium system dynamics*, that is, the change required within different elements of the system to ensure social and ecological resilience. Our collection of studies demonstrates that the *inclusion* of *community priorities and diverse forms of local knowledge* have the power to help farmers adapt to and prepare for changes, both in governance frames (like devolution) and in economic and environmental contexts (global price trends, climate change).

Assessment of the resilience umbrella tool

Like an umbrella, one does not wish to remain constantly and only under the cover of resilience. When we use umbrellas, we are waiting for the rain to end. Resilience might be seen as what an umbrella is at best: preparedness in case of storms, a readiness that is not relentlessly clutched under the arm, but maintained in good condition, and most times stored, ready for use, on a hook in a pantry full of the family's food. Farming-system resilience is a state of preparedness for and acceptance of change among citizens and political and economic decision-makers alike.

The Resilience Umbrella conceptual tool is a contribution to scholarly debate, scientific discovery and agricultural policy analysis, insofar as it illuminates distinct sites of social change within the constituent elements of the farming system, and therein enquires into social and ecological dynamics and direction of change. This analysis associates socio-ecological resilience in semi-arid farming systems with more inclusive access, more adaptive resource-use, more equitable livelihood options, and policies that need to be transformed to support the nutritional priorities and very local market systems that drove success in the project's achievements and results. Bahadur et al.'s (2013) characteristics of resilience, and Njuki and Miller's (2013) gender-transformative approach strengthen the Resilience Umbrella's analytical usefulness, by helping to specify the content and direction of changing gendered power relations within the farming system.

Questions for further study

Resilience in the farming system means *retaining all of the key elements* of the system (relationships, activities, policies, markets), and *maintaining the capacity of the system* as a whole to function as a resilient and enduring source of household food and nutrition security. To be resilient, this system must continue to function *at the same time* that relationships of actors within the system undergo change towards diversity, equity and inclusion. Change can be planned or unexpected. This collection has emphasized planned and intentional transformation of farming practices and gender relations, through research for development that tried to catalyze practical and lasting change in farmers' lives. The research in many ways took into account the environmental changes that the people of Eastern Kenya are experiencing. But the resilience framing we adopted has not yet been employed to analyse changes that come about due to sudden or catastrophic shocks, whether political, climatic, economic, or otherwise. The extent to which the tool is useful in these situations has yet to be explored.

It is not lost on us that while women of Eastern Kenya repeatedly shared with us their preference for the production of food crops to boost household food security, Nobel Laureates have also determined child nutrition to be the first Sustainable Development Goal to be achieved by 2030. Long-term endurance of nutritional security could potentially be built on the improvements of child health that are possible with widespread farmer adoption of the gendered technologies and practices detailed in the studies collected here. Further gains could follow as the children mature with better nutrition, enabling them to reach their full capacities. The gains to be made are incalculable, insofar as they centre on the health, life chances and human development of entire generations. If one-third of Kenyan children are malnourished to the point of stunting, as they are now, then every child in Kenya remains exposed to the insecurity of person and risks to health, peace and the environment that accompany such destabilizing chronic poverty. Thus whatever is invested in farming system resilience to elevate the country's children out of danger of starvation and stunting, improves not only the lives of that already substantial proportion of the population, but indeed the lives of every Kenyan.

The women small-scale farmers with whom we worked placed priority on their families' subsistence and household food sufficiency. For us, women farmers' preference and capacity to directly produce nutritious and locally adapted foods to feed their families constitute the starting place for measuring the success of development research, policy and practice. This standpoint recognizes and valorizes the food security contributions of small-scale farmers' subsistence priorities and practices. From this perspective, socio-ecological resilience will be most effectively advanced in semi-arid Kenyan farming systems by the coordinated efforts of research, policy and farmers, guided by the innovations prioritized by those most affected by hunger and malnutrition: Kenya's small-scale and subsistence women farmers.

Endnotes

1 There has been debate over what *kind of value* is counted in the term *high-value traditional crops* (HVTCs). Some see the nutritional value, others the cultural and ecological values of indigenous food-crop varieties grown, sorted, saved and shared by small-scale and subsistence farmers. These are non-priced values in addition to the monetary value of these crops, which others prioritize as the main value of HVTCs.

References

Akuku, B., Makini, F., Wasilwa, L., Makelo, M., and Kamau, G. (2014). *Application of Innovative ICT Tools for Linking Agricultural Research Knowledge and Extension Services to Farmers in Kenya*. Presented at the UbuntuNet–Connect Conference, Infrastructure, Innovation, Inclusion, in Lusaka, Zambia, 10–14 November 2014.

Bukania, Z. N., Mwangi, M., Karanja, R.M., Mutisya, R., Kombe, Y., Kaduka, L.U., and Johns, T. (2014). Food insecurity and not dietary diversity is a predictor of nutrition status in children within semiarid agro-ecological zones in Eastern Kenya. *Journal of Nutrition and Metabolism, vol. 2014*, article ID 907153.

Bahadur, A. V., Ibrahim, M., and Tanner, T. (2013). Characterising resilience: Unpacking the concept for tackling climate change and development. *Climate and Development, 5(1)*, 55–65.

Elson, D. (1991). *Male bias in the development process*. Manchester, UK: Manchester University Press.

Gurung, B., Ssendiwala, E., and Waithaka, M. (Eds.). (2011). *Influencing Change: Mainstreaming Gender Perspectives in Agricultural Research and Development in Eastern and Central Africa*. Cali, Colombia: International Center for Tropical Agriculture; and Entebbe, Uganda: Association for Strengthening Agricultural Research in Eastern and Central Africa.

Government of Kenya. (2011). NALEP within the policy framework. *Biashara Leo*, April/May.

FAO. (2011). *The State of Food and Agriculture: Women in Agriculture, Closing the Gender Gap for Development*. Rome, Italy: FAO.

FAO. (2013). *The State of Food and Agriculture: Food Systems for Better Nutrition*. Rome, Italy: FAO.

Kenyatta, M. (2015, June 9). Political leaders key in ending malnutrition. *The Standard*, http://www.standardmedia.co.ke/article/2000165010/political-leaders-key-in-ending-malnutrition

Meinzen-Dick, R., Quisumbing, A.R., and Behrman, J.A. (2014). A system that delivers: Integrating gender into agricultural research, development, and extension. In A.R. Quisumbing, R. Meinzen-Dick, T.L. Raney, A. Croppenstedt, J.A. Behrman, and A. Peterman (Eds.), *Gender in Agriculture: Closing the Knowledge gap* (pp. 373–391). Netherlands: Springer, Rome: FAO.

Njuguna, E., Brownhill, L., Kihoro, E., Muhammad, L. W., and Hickey, G. M. (in press). Gendered technology adoption and household food security in semi-arid Eastern Kenya. In J. Parkins, J. Njuki, and A. Kaler (Eds.). *Towards a Transformative Approach to Gender and Food Security in Low-Income Countries*. London, UK: Earthscan.

Njuki, J., and Miller, B. (2013). Making livestock research and development programs and policies more gender responsive. In J. Njuki and P.C. Sanginga (Eds.). *Women, Livestock Ownership and Markets: Bridging the Gender Gap in Eastern and Southern Africa*, (pp. 111–128). London, UK: Earthscan.

Ragasa, C., Sengupta, D., Osorio, M., Haddad, N.O., and Mathieson, K. (2014). *Gender-Specific Approaches, Rural Institutions and Technological Innovations*. Rome, Italy: FAO.

Sanginga, P. C., Waters-Bayer, A., Kaaria, S., Njuki, J., and Wettasinha, C. (Eds.) (2009). *Innovation Africa: Enriching Farmers' Livelihoods*. London, UK: Earthscan/Routledge.

Seavoy, R. E. (2000). *Subsistence and Economic Development*. Westport, CT: Praeger.

Staudt, K.A. (1978). Agricultural productivity gaps: A case study of male preference in government policy implementation. *Development and Change. 9*(3), 439–457.

Index